Frederick W. Bedford

Supplement to Third Edition of History of George Heriot's

Hospital

And the Heriot Foundation schools

Frederick W. Bedford

Supplement to Third Edition of History of George Heriot's Hospital
And the Heriot Foundation schools

ISBN/EAN: 9783337309008

Printed in Europe, USA, Canada, Australia, Japan

Cover: Foto ©berggeist007 / pixelio.de

More available books at **www.hansebooks.com**

SUPPLEMENT TO THIRD EDITION

OF

HISTORY

OF

GEORGE HERIOT'S HOSPITAL,

AND THE

HERIOT FOUNDATION SCHOOLS,

BY

FREDERICK W. BEDFORD, LL.D.,

HOUSE-GOVERNOR AND HEAD-MASTER OF HERIOT'S HOSPITAL,
AND INSPECTOR OF HERIOT FOUNDATION SCHOOLS.

EDINBURGH:

BELL & BRADFUTE, 12 BANK STREET.

1878.

PREFACE TO SUPPLEMENT TO THE
THIRD EDITION.

THE latest (Third) Edition of the History of Heriot's Hospital was published in July 1872. As the sale of this Book is limited principally to persons connected with the Institution, it requires several years to exhaust a moderately-sized Edition. Thirteen years elapsed between the Second and Third Editions. As the Text and Appendixes are frequently found by the Governors and Officials convenient for purposes of reference, the Editor has thought its usefulness will be much increased, if, between the issue of two Editions, a small Supplement be published containing a brief record of the principal events in the History of the Hospital since the publication of the previous Edition, so that the Book may never be more than a few years out of date. The present Supplementary pages have been prepared with this view.

Since 1872, six Out-door Schools (Three Juvenile and Three Infant) have been erected, and one temporarily established in the Fountainbridge District; Free Evening Classes have been formed for the instruction of Males and Females during the Winter Months; the important Burgess Act of 1876 has been passed ; Day Scholars have been admitted into the Hospital by competition from the Out-door Schools, and several other matters of interest to the Institution have occurred.

The Appendix contains a List of the present Governors, Officials, and Teachers ׃ . the Hospital and Schools; a Table of all the Heriot Schools, with the date of their opening and the amount of accommodation in each ; the Names of all the House-Bursars, Out-Bursars, Medallists, and Anniversary Preachers since 1872; the number of Boys annually admitted into the Hospital, and the Revenue and Expenditure of the Institution since that date. The amended Regulations, adopted in 1876, for the Management of the Heriot Schools are inserted in full.

For the Collection and Notes in the Appendix regarding the Trabroun Family, the Editor is indebted to Mr G. W. BALLINGALL, Taap Hall, Ferry Road, Edinburgh, who states that he obtained some items of useful information from Mr G. HERIOT STEVENS of Gullane, and J. RONALDSON LYELL, Esq., Lochy Bank, Auchtermuchty.

FRED. W. BEDFORD.

GEORGE HERIOT'S HOSPITAL,
February 1878.

CONTENTS.

SUPPLEMENT TO HISTORY

OF

GEORGE HERIOT'S HOSPITAL.

AT a General Meeting of the Governors of Heriot's Hospital on the 4th of April 1872 the Education Committee recommended that as there was then at the credit of the Schools Account a surplus of about £3,836, the Governors should remit to the Education and Finance Committees to consider the practicability of establishing an additional Out-door School or Schools, and also (seeing that there was no immediate prospect of any changes being made in connection with a private Act of Parliament) to consider and report as to charging fees in the Schools and Elementary Evening Classes; and further, to consider the salaries of various Teachers in the Hospital. The Joint-Committee to whom these matters were remitted reported on the 2d of May their opinion that if the Teachers were fairly entitled to an increase of their salaries their cases should be considered before the Governors should proceed to dispose of their surplus revenue by establishing additional schools; and, entertaining this view, they carefully considered the salaries, work, and services of the various Teachers,

a

and afterwards made recommendations that resulted in increased remuneration to all the principal Teachers in the Hospital and Schools, and subsequently to other officials of the Institution. This portion of the remit having been disposed of, the Joint-Committee, in August of the same year, recommended that the Governors should resolve, so far as arrangements could be made, that *four additional Out-door Schools should be established*, as, from a statement made by the Treasurer of the available means likely to accrue in the course of four or five years, the necessary expenditure for this undertaking would be fully warranted. It was suggested that the new Schools should be in the following localities—one at Fountainbridge, another on the South Side of the town; a third at Abbeyhill, and a fourth at Stockbridge, or elsewhere where most needed;[1] and it was remitted to a Sub-Committee to look out for and report on sites for the three first-mentioned Schools, as well as to consider and report where the fourth one should be set down, and to consider and report whether accommodation could in the meantime be rented in the districts referred to, so as to allow of temporary Schools being opened before permanent ones could be erected or acquired. The Governors resolved in terms of these recommendations, the understanding being that the fourth new School should be erected in Stockbridge. It was subsequently decided that temporary Schools should be established for Fountainbridge in the Ponton Street Hall; for Abbeyhill, in a large brick building belonging to Mr Nicolson; and for the South Side, in the Arthur Street United Presbyterian School. Three handsome permanent buildings, situated respectively in Abbeyhill, Davie Street,

---

[1] Record of Heriot's Hospital, vol. xliii. p. 195.

and Stockbridge, have since been erected from designs prepared by Mr John Chesser, Architect to the Hospital. Each of these buildings contains both a Juvenile and an Infant School, fitted up internally with the most modern appliances, and exhibiting externally some of the distinctive features of the grand old parent institution.

On October 3d, in the same year, the Governors approved of a recommendation made by the same Committee that Evening Classes be opened in three of the Outdoor Schools for instruction in Writing and Book-keeping, Arithmetic, Mathematics, English Composition and Literature, Architectural and Mechanical Drawing, Experimental Physics, French, and German, and Teachers for all these departments of instruction were subsequently appointed. The rate of fee was made the same as at the Watt Institution and School of Arts, viz., Five shillings for admission to a single Class, and two shillings and sixpence for every additional class. This arrangement was in successful operation during the Session 1872-73.

Meanwhile the Governors had under consideration the important question whether fees should be charged to the pupils attending the Out-door Day Schools, and it had been resolved by the Governors on the 3d of October 1872 to remit to the Law Committee with instructions to take the opinion of Counsel on the legality of charging school-fees. At the General Meeting on 11th November of the same year, the Clerk laid on the table the opinion of the Solicitor-General (Mr Rutherfurd Clark) and Mr Wm. Watson, on the following Queries submitted to them by the Law Committee :—

" 1. Have the Governors, under the Act of 1836, power to make a rule or regulation whereby the children attending their Out-door Schools shall pay a small fee for such atten-

dance, either for the advantage generally which they are to
enjoy, or in respect of the books, pens, ink, and paper,
&c., with which they are supplied?

" 2. Or are the sons of Burgesses or Freemen entitled to
be exempt from any such fee?"

The following is the opinion of Counsel:—

" 1 and 2.—We are of opinion that the Memorialists have
no power to charge fees, however small in amount, for
teaching in their Schools any child or children belonging to
either of the three classes enumerated in section 5 of the
Act of 1836.

"It does not seem to admit of doubt that the leading
purpose of the foundation of the Hospital, as settled by the
will of the founder and the statute of Dr Balcanquall, is
to afford free education and maintenance to poor scholars,
children of Burgesses and Freemen of the City of Edinburgh.   The foundation as thus settled was modified by
the Act of 1836; but it appears to us that, according to
the fair construction of the provisions of that statute, these
were intended, not to alter the character of the Institution,
put to extend its scope by giving to a wider class of
recipients, through the medium of Out-door Schools, the
benefits of a free education in accordance with the true
spirit of the 'Pious Donation.'

" It may be made matter of doubt whether the
Memorialists, besides furnishing School accommodation and
the services of Teachers, are bound to provide books, pens,
ink, and paper, for the use of children entitled to the
benefit of being taught in their Schools. We are inclined to ·
hold that it is incumbent upon the Memorialists to do so.
But, apart from any such questions, we are of opinion that
the Memorialists are not entitled to charge fees for the use

of books, paper, &c., belonging to the Hospital; and that, even if the parents or guardians of the children were bound to provide these articles, the Memorialists would not be entitled to require that such articles shall be purchased from themselves.

" We venture to suggest, that if the Memorialists desire an authoritative opinion in regard to their right to charge fees, that object may be easily attained by means of a Special Case. Considering the difficulty and importance of the question, it appears to us that the Court would ordain the whole expenses of such Special Case to be charged against the revenues of the Hospital." [2]

The Governors having considered the Case and Opinion, remitted the same to the Education Committee, with power to consult the Law Committee thereon, and, if considered necessary, to take the opinion of Counsel as to the right of the Governors to charge fees for the Pupils attending the Evening Classes which had been opened in the Schools.

The Education Committee reported in September 1873, that as the Evening Classes appeared to have been highly successful the previous Winter, it would be very desirable that similar Classes should to a certain extent be held the following Winter, although it seemed doubtful, from the opinion of Counsel given in October 1872, whether the Governors are entitled to charge fees for attendance. It was therefore proposed that the Teachers to be appointed should have the free use of the school furniture and apparatus, with fire, gas, and cleaning, that a sum not exceeding £10 should be payable by the Governors to the Teachers of the Elementary Classes, together with a fee of 2s. 6d. payable by the pupils in attendance ; that a fee of 5s. should be payable by each pupil to the Teachers of the Special Classes, but no salary from the Governors.

[2] Record of Heriot's Hospital, vol. xliii, p. 232.

The Governors resolved, after considerable discussion, that this recommendation of the Committee be re-committed for further consideration, and on 1st November 1873 it was further resolved that, out of deference to opinions expressed at previous meetings, a trial should be made of *free* instruction in all the Classes, both Special and Elementary, and that the pupils admissible to the Classes be those referred to in Section 5 of the Hospital's Act of Parliament authorising the establishment of Out-door Schools, and in the order of preference therein mentioned. It was also decided that the Elementary Classes should be opened and closed with prayer, and that a small portion of the Bible be read at the close, and also that the scholars attending the Elementary Classes might have the use of books employed in the Schools. These arrangements have been in operation ever since. The Classes have been annually reported upon by Examiners appointed by the Governors, and Certificates are awarded to such of the scholars as have made a certain proportion of attendances, and attained a satisfactory standard of proficiency.

On the 15th of April 1873 the Governors and the Institution sustained a severe loss in the death of Mr Isaac Bayley, the Clerk and Law Agent of the Hospital, who for nearly forty-three years had occupied that office. At the meeting of the General Board on the 5th of June, the Rev. Dr Stevenson, on behalf of a Sub-Committee appointed on the suggestion of the Lord Provost, submitted the following Minute :—" This being the first Ordinary Meeting of the Board since the lamented death of their Clerk Mr Isaac Bayley, the Governors resolved, before entering on the business, to record in their Minutes an expression of the deep and lasting gratitude which they feel to be due to his memory for the probity and prudence with which, both at the Board and in the conduct of its various Committees, he ever discharged the duties of his office, and of the high esteem in which he was held by the Governors both as a Christian and a gentleman. His intimate acquaintance with the constitution and history of the Hospital, and the warm and unwearied interest which he took in its affairs, inspired the minds of the Governors with great confidence in the wisdom of his advice on all occasions, and contributed largely to the prosperity and usefulness of this great Institution. The Governors accordingly feel that by his death they have sustained the loss of a true friend as well as of a much valued officer, and in token of their sympathy they direct a copy of this Minute to be conveyed to the surviving members of his family." The Governors unanimously approved of this Minute, and directed it to be entered upon the Records, and requested the Treasurer to transmit a copy of it to the late Mr Bayley's family.

Mr George Bayley, W.S., who for many years previously had been a partner with his Father, and had during that time assisted him in the business of the Hospital, was unanimously appointed Clerk and Law Agent in room of his late Father.

Mr William Forrester, the Treasurer of the Hospital, having by letter, dated 18th September 1873, intimated his desire, from his age and the increasing duties devolving upon him, to resign his office, and to be relieved of duty by the beginning of October, the Governors, at their Meeting on the 18th of September, remitted this letter to a Special Committee, that they might consider and report thereon along with a remit as to the duties and emoluments of the office.

The Governors marked their sense of the valuable services that Mr Forrester had for thirteen years rendered to the Hospital by asking his acceptance of the gift of one year's salary (then £500), along with a piece of plate having a suitable inscription engraved upon it ; and on the 2d of October they unanimously approved of the following Minute as expressive of their feelings on the occasion, directed the same to be engrossed in the Records of the Hospital, and authorised the Clerk to transmit a copy thereof to Mr Forrester :—" In accepting the resignation of Mr Forrester, Treasurer of the Hospital, the Governors desire to record the high sense they entertain of the value of his services, and the manner in which he has discharged the duties during the thirteen years he has held the office. His previous acquaintance with the management of the Institution, and his knowledge of business, combined with his faithful and unwearied attention to every detail in the pecuniary affairs of the Trust, have contributed much to its prosperity and success, whilst his quiet and kindly manner have won the regard and esteem of all—Governors, Officials, Teachers, and Pupils alike—with whom he has been brought into contact in the discharge of the many, varied, and responsible duties devolving upon him. They earnestly trust that every blessing may attend him in his retirement, and that he may be long spared to enjoy the friendship he has formed in his official connections, and to see the Institution in which he is so interested continuing to be a blessing to the public of the city."

On the 6th of October 1873 Mr David Lewis, one of the Magistrates of Edinburgh, was unanimously appointed to the office of Treasurer, in place of Mr Forrester resigned.

On the 2d of October 1873 the Governors decided to allow the admission of Sewing Machines into the Schools, the understanding being that such children only be allowed to use them as have become proficient in plain needle-work.

On 23d January 1874 the front of the Hospital was tastefully illuminated on the occasion of the Marriage of H.R.H. the Duke of Edinburgh to the Grand Duchess Marie of Russia, as had previously been done at the Marriage of the Prince of Wales in 1863.

On March 29, 1874, the death was reported of Mr David Simson, who for nearly thirty-nine years had been Drawing Master in the Hospital. Only two months previously Mr Simson had been compelled by the infirmities of age to resign his appointment. The Governors had, together with a highly complimentary reference to his past valuable services, granted him a retiring allowance of half of his salary, and on the intimation of his death they further resolved to present his representatives with the amount of one year's retiring allowance.[1]

At the Meeting of the Governors on 15th of February 1875 the Governors remitted to the Clerk and Law Agent to prepare a Memorial and lay it before Counsel for their opinion as to the right of widows who are Burgesses to have their sons elected to the Hospital. The following Memorial and Queries were laid before Mr W. Watson, then Solicitor-General (afterwards Lord-Advocate), and Mr W. E. Gloag :—

---

[1] Mr Simson was a well-known and highly-respected citizen. The following lines to his revered memory appeared in the *Scotsman* on the day after his funeral :—

"Lay him down softly in the mother's lap,
　No gentler spirit e'er has left the day;
Ah, yes! ye flowers now bright with vernal bloom,
　Who loved you comes to rest with you for aye!

"So smile your brightest o'er his grassy bed,
　And with your sweetest perfume fill the air;
For only then will ye be like to him,
　Whose life was fragrant with a sweetness rare!"

H. G. C. S.

" 1. Are the terms 'Burgesses and Freemen' synonymous terms, or could, or can now, a person be a Burgess without being also a Freeman, or *vice versa?*

" 2. Can females be legally elected Burgesses and Freemen of the Burgh of Edinburgh ?

" 3. Assuming Query No. 2 to be answered in the affirmative, are the sons of all female Burgesses and Freemen without distinction eligible for admission to Heriot's Hospital, or is a distinction to be made—

> "(1) In the case of sons of females who became Burgesses and Freemen before they were married ?
>
> "(2) In the case of illegitimate sons of unmarried female Burgesses ?
>
> "(3) In the case of the sons of females who became Burgesses and Freemen after they were married, their husbands not being or ever becoming Burgesses and Freemen ?
>
> "(4) In the case of sons of widows who became Burgesses and Freemen after their husbands' death, their husbands not having been Burgesses and Freemen ?

" 4. Assuming Counsel to be of opinion that females can be legally elected Burgesses and Freemen, and that their sons in all or any of the cases mentioned under the 3d Query are eligible for admission to the Hospital, are the sons of widow Burgesses and Freemen (their husbands not having been Burgesses and Freemen) entitled to the preference conferred upon fatherless boys, sons of Burgesses and Freemen, by the Hospital's Act of Parliament ?

" 5. Are illegitimate sons of Burgesses and Freemen eligible for admission to the benefits of the Hospital ?

" Counsel are requested to favour the Memorialists with

their opinion on any points arising out. of the Memorial
which are not brought out by the preceding Queries."

On the 30th of July in the same year the Clerk submitted the following Opinion :—

" 1. We are of opinion that, according to a sound construction of Heriot's Disposition and Assignation, his last Will,
the Statutes of Dr Balcanquall, and the Act 6 and 7 George
IV., c. 15, the meaning of the word Burgess is not restricted
by the use of the term Freeman in conjunction with it, and
that the benefits of the Foundation are conferred on the
sons of Burgesses and of Freemen being Burgesses. We do
not think that the benefit was restricted to those Burgesses
only who were members of Incorporated Trades or of the
Merchant Company, but was extended to all Burgesses
alike.

" 2. We think this question involves a point of very great
difficulty, but, on the whole, we are disposed to answer it in
the negative. To make a female a Burgess, to the effect at
least of conferring on her children the benefits provided by
George Heriot's Will, seems to us to be inconsistent with
the tenure of Burgess holdings and the understanding and
practice of the country. It appears to us that at the time
when George Heriot lived there were no female Burgesses in
Scotland, and that he could not have contemplated in objects
of his charity any others than the sons of Burgesses who
were males. It is brought out in the Memorial that the
only object now sought to be attained by conferring on
females a Burgess Ticket is to enable them to participate in
the benefit of Heriot's Hospital, and thus in our opinion to
direct the Funds of the Hospital to individuals who were
not within the contemplation of the Trust. It appears to
us that the Will of the Truster cannot be lawfully defeated

in this manner. The further question, How far the Memorialists are entitled or bound to refuse effect to a Burgess's Ticket given by a Magistrate, is attended by considerable difficulty, but in our opinion they ought to endeavour to carry out the directions of the Truster according to their true scope and intent, and to refuse to acknowledge female Burgesses until their right to participate in the benefits of the charity is established by competent authority.

"3. It follows from our answer to Query 2, that, in our opinion, the sons of females admitted as Burgesses are not eligible. It may be proper, however, to say that had we been of opinion that the Memorialists were bound to recognise the position as Burgesses of females, we should have thought that the children of all such females holding a burgess-ticket were eligible, and we could not have drawn any distinction between the sons of females in the different positions mentioned in the sub-division of this query, if the poverty of the parents were established, excepting that we are of opinion that the illegitimate sons of unmarried female Burgesses would not in any event be eligible.

"4. We are of opinion that in a sound construction of the clause of the statute here referred to, the preference thereby created is conferred only on children whose fathers are dead and were Burgesses, and that the son of a widow admitted as a Burgess, but whose father was not a. Burgess, is not entitled to claim the preference.

"5. We are of opinion that the Deeds and Statutes regulating the management of Heriot's Hospital, do not refer to illegitimate children, and we therefore answer this question in the negative.".

The Governors having considered the Minutes and

Report, with the Queries and Opinion of Counsel, determined by a large majority "That the Governors resolve, in accordance with the opinion of Counsel, not to elect sons of female Burgesses and Freemen to the Hospital."

In 1876 a Bill[1] was introduced into Parliament by Mr Duncan M'Laren, M.P. for Edinburgh, "to assimilate the Law of Scotland to that of England as regards the Creation of Burgesses." This Bill, which received the Royal assent on 1st June 1876, was at once seen to be, as by the introducer of the Bill it was professedly designed to be, of the greatest importance to Heriot's Hospital. The number of Burgesses

---

[1] The portion of the Bill in which Heriot's Hospital is interested is as under:—

"Whereas an Act was passed in the 5th and 6th year of His Majesty William IV., chap. 76, intituled 'An Act to provide for the Regulation of Municipal Corporations in England and Wales:'

"And whereas another Act was passed in the 32d and 33d year of Her Majesty, chap. 55, amending the same:

"And whereas it is expedient to assimilate the law of Scotland in some respects to the law of England as regards the Creation of Burgesses:

"Be it therefore enacted by the Queen's Most Excellent Majesty, by and with the advice and consent of the Lords Spiritual and Temporal and Commons in this present Parliament assembled, and by the authority of the same:

"1. Every person in Scotland of full age, liable to be rated for the relief of the poor, who at the term of Whitsunday 1876, or any succeeding term of Whitsunday in any year, shall have occupied any house, warehouse, counting-house, shop, or other building, within any burgh in which there are Burgesses, during the whole of that year and the whole of the two preceding years, and who during the time of such occupation shall have been an inhabitant householder within the said burgh, and who shall have been rated in respect of such premises so occupied within the burgh to all rates made for the relief of the poor of the parish wherein such premises are situated during the term of his occupation as aforesaid, and who shall have paid, on or before the last term of Whitsunday as aforesaid, all such rates, together with all burgh rates, if any, as shall have become payable in respect of the said premises, except such as shall have been payable within six calendar months next before the said last term of Whitsunday, shall be, subject to the conditions herein-after contained, a Burgess of such burgh so long as such person shall occupy

before that date was believed not to have exceeded 500, but by the terms of this Bill the number that immediately became entitled to the privileges of Burgess-ship seemed, in the opinion of many, to have been not less than 30,000. It became immediately obvious that the relation of Heriot's Hospital to the public of Edinburgh would be altered to an important extent. If there had been any doubt in former years as to the necessity or expediency of such an Institution for the relief of the limited number of citizens' children whose fathers' premature death or unsuccess in business had

premises and be rated and pay rates in manner aforesaid within the same : Provided that the premises in respect of the occupation of which any person shall have been so rated need not be the same premises or in the same parish, but may be different premises in the same parish or different parishes : Provided also, that no person being an alien, and no person who within twelve calendar months next before the last term of Whitsunday shall have received parochial relief, or any pension or charitable allowance from the Town Council revenues of such burgh, or from any corporate body within the same, shall by virtue of this Act be held to be a Burgess of such burgh so long as he continues to receive such pension or charitable allowance : Provided further, that no person shall be disqualified from ,being a Burgess as aforesaid by reason that any child of such person shall have been admitted and taught within any Endowed School.

" 2. Nothing herein contained shall interfere with any law or legal usage by which Burgesses are now created or admitted in any burgh, or give or imply any right or title to or interest in any merchant's house or trades' house, or any patrimonial lands, common or other properties, funds or revenues of any of the Guilds, Burgesses of Guilds, Crafts, or Incorporations of the burgh, or to or in any Burgess acres or any grazing rights connected there- with, or any mortifications or benefactions for behoof of the members of such Guilds, Burgesses of Guild, Crafts, or Incorporations, or of their families, or any right of management thereof, or any membership in any of the said Guilds, Burgesses of Guild, Crafts, or Incorporations, or of such Burgess acres : Provided that the widows and children of Burgesses admitted under this Act, and who may die during the period of their Burgess-ship, shall have and enjoy all the rights and privileges which the widows and children of simple Bur- gesses created or admitted in any other manner now enjoy by the law and practice of Scotland."

left their families in comparative destitution, it was felt that out of the families now entitled in their misfortunes to become applicants for Hospital benefits a sufficient number of thoroughly needy and deserving cases could be found to justify its continued employment as a Home and "Seminarie of Instruction." Moreover, in the prospect of the introduction of a graded system of instruction for the benefit of children attending the Heriot Schools, the important fact that Burgess-ship in Edinburgh would become henceforth practically identical with citizenship, held out the hope that the son of the poorest ratepayer might henceforth become entitled to the highest educational privileges.

Ever since the Home Secretary's rejection, in 1871, of the Provisional Order then applied for by the Governors, there had been a desire to take advantage of any favourable opportunity for making such important changes in the Hospital as "altered circumstances render expedient," and one of the first practical results of the passing of the Burgess Bill was the following Motion, proposed by Bailie Tawse, and adopted by the Governors on 21st August 1876 : —"That with the view of promoting the higher Education of those attending George Heriot's Hospital Schools, there shall, as soon as possible, be elected and admitted to the Hospital as Scholars twelve or such other number of boys, sons of Burgesses of the City of Edinburgh, (fatherless boys being always preferred) who have attended one of said schools, and such scholars shall be entitled to attend the Hospital classes for their Education, and while so attending (not being above 16 years of age, nor obtaining a Bursary) they shall each be allowed in lieu of maintenance a sum of £5 annually, to be expended in clothing, it being declared that such scholars shall be elected after a competitive examination and shewing a proficiency not under that

required for a successful pass in Standard   of the
Code of 1876, and that it be remitted to the Education
Committee to frame all necessary rules and regulations for
carrying out said Resolution and conducting the examination,
and also to intimate this resolution and all further particu-
lars to the different schools."

In accordance with this decision of the Governors, 12
sons of Burgesses, educated in the Out-door Schools, were
elected as scholars into the Hospital after a competitive
examination on the 19th of February 1877, and 12 more,
similarly qualified, were admitted into the Hospital six
months afterwards. In November 1877 the Governors
agreed to admit 6 more in February 1878. The total
number of boys then receiving education in the Hospital
will be—120 Residents, 60 Non-Residents, and 30 Day-
Scholars.

In November 1876 the Treasurer submitted to the
Endowed Schools Committee a formal representation, in
which he reminded them of the published Report of the
Endowed School Commissioners, which had not yet been
disposed of by the Legislature, quoted the views expressed by individual Members of the Commission at
various meetings of the recently established "Association
for promoting Secondary Education," and suggested that as,
in his opinion, the interests of the Hospital were likely by
delay to be placed in imminent peril, the Committee should .
consider whether it ought not to report to the Governors at
their next General Meeting what course of action would
be desirable for the purpose of protecting the funds of the
Heriot Trust, and extending its benefits to that class whose
relief was primarily contemplated by the Founder. The
Committee having considered this representation and
the whole circumstances, resolved to recommend to the

Governors to apply to Parliament in the ensuing Session for a Private Act for the purpose of obtaining powers to extend the benefits of the Foundation in the same directions as were indicated in the proposed Provisional Order of 1870, and with that view the Committee instructed the Clerk to prepare Notice of a Bill for publication. The Governors having considered this Recommendation and Notice, the following motion, proposed by the Rev. Dr Gray, was, after a long discussion, unanimously agreed to, and the Governors remitted in terms thereof:—"The Governors, after full consideration of the recommendation of the Special Committee, before coming to any decision as to the introduction of a private Bill, remit to the Sub-Committee, with the addition of Rev. Mr Giffen and Rev. Dr Stevenson, to communicate with the Lord Advocate, and, if necessary, with the Home Secretary, to ascertain whether the Government are prepared to introduce a Bill empowering the Governors to prepare a Provisional Order which will enable them to extend the benefits of the Heriot Trust to the class of beneficiaries now receiving the advantages of said Trust." The Special Committee, having had an interview with the Lord Advocate, reported that his Lordship possessed no information which would guide the Governors in coming to a decision on this matter, whereupon Councillor Anderson, seconded by Councillor Harrison, proposed the following motion:—"That while the Governors are of opinion that some extension of the powers they now possess is needed to enable them to extend the benefits of the Heriot Trust more widely among the large number of Burgesses created by the recent Burgess Act, they do not consider it expedient to adopt the recommendation of the Special Committee, to apply for a Private Act in the next

B

Session of Parliament, but resolve to re-appoint the Committee, with instructions to watch generally over the
interests of the Trust, and specially to communicate with
the Home Secretary as to the Reports of the Endowed
Schools Commission and the intended acts of the Government in regard to it, and to endeavour to procure a General
measure of Parliament under which powers may be
given to effect all necessary reforms by means of a Provisional Order, or otherwise ; further, instruct the Committee to reconsider as to whether effect can now be given
to any of the unexercised powers possessed by the Governors, and to report as to them as soon as possible."

The following motion was thereafter proposed by Bailie
Cranston and seconded by Councillor Gowans :—"The result
of the Burgesses (Scotland) Act 1876, as applicable to Edinburgh, having been to increase the Burgess Roll, qualifying
for admission to the Hospital, from about 500 to upwards of
24,000, it is necessary, and the Governors resolve, to build
new schools for the secondary education of sons and
daughters of Burgesses who may evince qualities encourag-
ing higher education ; also to assist such scholars with such
pecuniary maintenance as the Governors deem necessary ;
and further to aid such of the scholars as proceed to the
University ; and in order to effect these objects the Governors
resolve to ask such Parliamentary powers in next Session as
may be necessary to enable them to apply the resources of
the Hospital."   After a long discussion, 22 of the Governors
voted for Councillor Anderson's motion, and 8 for Bailie
Cranston's.   A formal protest against this decision of the
Governors was lodged by the mover and seconder of the
unsuccessful motion, and also by Bailie Colston and Councillors Mossman and Buchanan.

A further effect of the Burgess Act of 1876 was a motion

by Bailie Cranston in February 1877 to rescind the decision of the Governors of 4th of August 1875, " That the
sons of Female Burgesses in their own right be not elected."
After discussion, a motion by Treasurer Wilson was carried,
to the effect "that the whole matter as to Female Burgesses,
and also questions which have been started as to the eligi-
bility of parties as Burgesses under the Burgess Act, 39
Vict. cap. 11 (1st June 1876) be remitted to the Law Committee to consider and take the opinion of counsel upon."

In accordance with this decision the following Memorial
was submitted to the same Counsel (Lord Advocate Watson
and Mr W. E. Gloag), as had been consulted on a similar
question two years previously :—

" In July 1875 the Governors obtained the Opinion of
Counsel, *inter alia*, upon the query—Can Females be
legally elected Burgesses and Freeman of the Burgh of
Edinburgh ? The opinion of Counsel was—' We think
this question involves a point of very great difficulty, but,
on the whole, we are disposed to answer it in the negatives.
Counsel further stated their reasons for arriving at this
opinion, and advised that the Governors should refuse to
acknowledge Female Burgesses until their right to participate in the benefits of the Charity (Heriot's Hospital) is
established by competent authority. The Governors have
acted upon this opinion and advice, and have not admitted
Female Burgesses to the benefits of the Hospital, and no
Female Burgess has taken steps to establish her rights as
such.

" The Memorial submitted to Counsel in July 1875, and
their Opinion thereon, is herewith sent in reference to the
further Opinion which the Governors now desire in reference
to the query before quoted. At p. 7 of that document it is
stated that from an examination which had been made of

the Record of Burgesses kept by the City, it appeared that
on 17th of March 1407 a female, Mivona de Duscoe, was
admitted a Burgess, but that it did not appear from that
record that any other female had been admitted a Burgess
until 1869, when, as explained in the Memorial, the Town-
Council thereafter elected Females as Burgesses.

"Recently, however, attention has been drawn to a work
entitled "Extracts from the Records of the Burgh of Edinburgh," A.D. 1403–1528, published in 1869 by Mr J. D.
Marwick, late City-Clerk in Edinburgh, in which reference
is made to the subject, and the attention of Counsel is requested to whether the extracts therefrom, after quoted, in
any way influence the result they arrived at by their former
opinion.   Besides this, since Counsel gave their Opinion,
the Legislature has, by the Burgesses (Scotland) Act, 39
Vict. cap. 11, provided that "every person in Scotland of
full age" who has paid certain rates, &c., &c., shall be a
Burgess, so that females undoubtedly now become Burgesses,
and without the intervention of, or being created Burgesses
by Town-Council.   In fact, females become Burgesses
whether they wish it or not, if they are under the provisions
of the Act.

"As Counsel in their opinion founded to a certain extent
upon the "understanding and practice of the country"
with regard to Female Burgesses, the Governors desire, in
view of the Extracts from the Record of the Burgh of Edinburgh, and in view also of the passing of the Burgess Act,
to bring the before-mentioned query again under the attention of Counsel, and request their opinion of new thereon, seeing that the Opinion formerly given was arrived at
with difficulty.[1]

---

[1] The passages in the Extracts from the Records of the Burgh of Edinburgh
before referred to are as follows :—

" The Governors have only further to ask the attention
of Counsel to the Burgess Act of 1876, before referred to,
and on consideration of it, and the extracts before quoted,

1st December 1450.

" The Burges air payes for his Burgess-ship vjs. viijd. He that buys his
burgess-chip payes xls. Ane Burges be richt of his wyfe payes vjs. viid.
At this time many ar maid Burgessis and call it postnati et postnate of their
fatheris."

Under same date there is another entry of which the following is a translation :—" David Adminby was made a hereditary Burgess in right of his
uncle John Crysteson."

30th October 1453.

Of this date there is an entry, of which the following is a translation :—
" John Heane junior is made hereditary Burgess in right of his uncle Thomas
Spens, and paid for the freedom spices and wines, Thomas Lavington being
surety."

From this entry it would appear that John Heane was admitted a Burgess
in right of his mother's brother.

7th October 1462.

Of this date there is an entry, of which the following is a translation :—
" John Chapellane made gild brother in right of his wife Agnes, daughter of
John White and compounded by paying twenty shillings."

Before Michaelmas 1472.

" Before Michalmas 1472 ane James Furde Burges be his wyfu for vjs. viijd."

29th January 1477-8.

Under this date there is an entry of which the following is a translation :—
" Thomas Haliburton made Burgess and guild brother in right of his wife, &c.,
and paid spices and wine, and for the gild twenty shillings.

23d March 1507-8.

" The qubilk day it is stated and ordanit that in tyme cuming the privi-
eges of burges barnes be observit and kepit in this wis that the burges eldest
son sall pay enterand as air to his fader sall pay for his Burgesry vjs. viiid.
and for gildry enterand be hes fader brother of the gild xiijs. iiijd. and for the
second son his Burgesry xiijs. iiijd. and for his gildry xxs. and siclike the

and the Memorial of 1875, and Counsel's Opinion thereon, to favour them now with their Opinion upon the following Query :—Is Counsel still of opinion that females could not, prior to the Burgess Act of 1876, be legally elected Burgesses and Freemen of the Burgh of Edinburgh, and is Counsel of opinion that females can become Burgesses only under the provisions of the Burgess Act of 1876.

"In the event of Counsel, in consideration of the additional information now laid before them, and specially in view of the altered circumstances brought about by the Act of 1876, coming to a different opinion on this query from that given on the Memorial of 1875, they are re-

---

Burges doichteris lauchfulle gotten to have the privilege of the second son, viz., for the Burgesry lxiijs. iiijd. and for the gildry xx. tss."

<hr>

4th December 1513.

It is statute and ordanit be the president baillies and counsale for the commoun weill of this guid town that the watche sall be nychtlie had of xxiiij personis sufficient, and thairfoir that all personis nychbouris of this toun sall furnish the samyn in their expenssis baith wedowis and others and "ilkane of them that disobeyis sall pay twa s for ane unlaw unforgevin and to be poyndit, thearfore als oft as the case happinis and that these that has disobeyit in tyme by gane pay for ilk tyme xiijd failye and thairof there to be poyndit for ij s ilk time."

<hr>

15th July 1539.

The quhilk day the provest baillies counsale and men of gude merchanttis for the tyme hes statute and ordanit that all maner of personis allegand or • pretendand them to be fre and comburgessis of this burgh in using of the privileges thareof compeir and enter theme selffs with xi. dayis nixt heireafter within the burgh to remain and mak residence thairintil scott lott waird and walk with the otheris nychbouris comburgessis of this burgh; certefeing thame that gif thay failze heirintil thay in all tymes after sall be reput and holdan as unfremen and als dischargit the said xi dayis being past of thaie privilege and fredume gyf they any have and tretit as efferis unfremen in tyme cuming; and thatoppin proclamatioun be made hereof at the market croc of this Burgh for thame that comperis nocht efter the said xi days.

quested to consider the 3d Query to that Memorial[1] and
favour the Governors with their opinion thereon."

The following is the Opinion of the Lord Advocate
Watson and Mr W. E. Gloag, given on 29th Nov. 1877:—

"We remain of opinion that females could not, prior to
the Act of 1876, be legally elected Burgesses and Freemen
of the Burgh of Edinburgh, to the effect of entitling their
sons to participate in the benefits conferred on sons of
Burgesses by George Heriot's will and the statutes of Dr
Balcanquall. We think that the passages quoted from the
Records of the Burgh relate chiefly, if not entirely, to the
privileged position conceded to the husbands of Burgesses'
daughters when entering as Burgesses. And we believe
that it is in accordance with the general, if not the uni-
versal, practice of Burghs and Guilds in Scotland to admit
the sons-in-law of Burgesses or Guild Brethren on more
favourable terms than strangers. Understanding it to be
the fact that from 1407 to 1869 no female has been entered
in the Burgess Roll of Edinburgh, the Excerpts made from
the Records in this Memorial do not affect our previous
opinion, that in the time of Heriot it was the understanding and practice of the country that all Burgesses should
be men, and none females.

"The terms of the Burgess Act of 1876 do not appear to
us to affect the opinion above expressed to any extent.
We do not express any opinion as to the effect of that Act
itself.

"It is unnecessary to say more as to Query 3 of the
former Memorial than that we adhere to the opinion
formerly expressed by us."

---

[1] See Memorial and Queries laid before Counsel as to rights of Widows
that are Burgesses, p. 10 of Supplement to History.

On 7th June 1877 Councillor Harrison gave notice of the following motion :—"That any child may be withdrawn by the parents from instruction in the Shorter Catechism in George Heriot's Schools, and remit to the Education Committee to give effect to this resolution." A few months previously an application had been unsuccessfully made to the Governors by a Unitarian father for his child to be exempted from instruction in the Shorter Catechism. A similar application [1] by a Roman Catholic mother in 1866, and another from a Jewess mother in 1857, had been also refused, on the ground of inexpediency on the part of some of the Governors, and of assumed illegality on the part of others. It was thought desirable that the opinion of Counsel should be taken on the question of legality, and the following Queries were therefore submitted to the Lord Advocate (Mr William Watson) and Mr J. B. Balfour :—

" 1. Is it compulsory on the Governors to have a Catechism taught to the children attending the Hospital's Outdoor Schools?

" 2. Assuming Query No. 1 to be answered in the affirmative, Have the Governors nevertheless power to allow parents who desire it to withdraw their children from instruction in the Catechism taught?

" 3. Assuming Query No. 1 to be answered in the negative, and that Counsel is of opinion that it is optional in the Governors to have a Catechism taught to the children, Have the Governors, in the event of their resolving to have a Catechism taught to the children, power to allow parents who desire it to withdraw their children from instruction in the Catechism taught?

---

[1] See Foot-note on p. 259 of History of Heriot's Hospital.

"Counsel will be pleased to favour the Memorialists with their opinion on any other points embraced within the Memorial which are not brought out by the preceding queries."

On the 29th of December 1877 the following Opinion was given by Counsel :—

" 1. We are of opinion that it is not compulsory to have a Catechism taught to the children attending the Out-door Schools. There is a marked distinction between the provisions of the Act of 1836, which empower the Governors to make new rules and regulations for the management of the Hospital, and to alter the code of statutes framed by Dr Balcanquhal, and those provisions which have reference to their powers of enacting rules for the Out-door Schools. By section 3 of the Act the Governors are empowered to establish regulations for the management of the Hospital, provided always that these regulations shall 'not be inconsistent with the statutes of Dr Balcanquhal, the alterations to be made thereon by this Act, the laws of Scotland, and form of Church government therein by law established.' The statutes in question expressly require that a Catechism approved by the ministers of Edinburgh shall be taught in the Hospital, and section 4 of the Act, which empowers the Governors to make certain alterations upon them, contains the important *proviso* 'that all such alterations shall be consistent with the law of Scotland and the form of Church government therein by law established.' But no .such restrictions are placed by section 5 upon the power thereby conferred on the Governors to determine the rules and regulations under which children are to be taught and educated in the schools. We think this distinction must be held to have been designed, and that the Legislature must be taken to have entrusted to the Governors the absolute discretion of determining what shall be taught in the schools, and under what regulations and conditions the teaching shall be conducted.

" 2. Even if the statute were to be construed as making it compulsory upon the Governors to have a Catechism taught in the Out-door Schools, we consider that its terms would still leave it within the power of the Governors to allow parents who desired it to withdraw their children from instruction in the Catechism.

" 3. We answer this question in the affirmative, for reasons which have already been sufficiently explained."

This Opinion of Counsel was submitted to a meeting of the Governors in January 1878, when a motion of Councillor Harrison's, similar in its terms to the one proposed in June 1877, was, after a brief discussion, adopted, and the Education Committee was instructed accordingly.

At the same meeting it was agreed, on the motion of Bailie Tawse, " That as the new schools at Stockbridge are now ready for occupation, it be remitted to the Education Committee to consider whether one of the (Out-door) Schools should not now be set apart for more extended instruction in higher or specific subjects, or for Secondary Education."

From the brief narrative of events recorded in these Supplementary pages, it will be obvious that there has been during the last few years an active endeavour on the part of the Governors to extend the benefits of the Institution, as opportunities occurred, to the utmost extent of their statutory powers and financial means. Further changes are considered desirable, but these it is thought cannot be made without increased powers from the Legislature or from the high legal functionaries who are empowered by the Hospital's Act of Parliament to sanction, under certain specified conditions, proposed changes in the Code of Statutes.

The importance of Heriot's Hospital as a great educational institution, extending its benefits to so many thousands of the citizens through its Hospital, Elementary Schools, and Evening Classes, is annually increasing, and it must be a source of unmixed gratification to Mr Duncan M'Laren, the venerable Member for the City, whose name and services must ever be regarded as supreme in the establishment of the Heriot Schools, that the far-sighted scheme initiated in 1836 has been gradually productive of such stupendous results.

APPENDIX.

APPENDIX.

I.—LIST OF BURSARS.

OUT BURSARS.

(Continued from page 377 of History of Heriot's Hospital.)

1872 Peter Dewar.	1875 John F. Andison.
... David Jamie.	... William Milne.
... John R. Hursat.	1876 George H. Boyd.
... Alfred Daniel.	... Harry Ranken.
1873 Alex. Smellie.	... Alexander Morton.
... James M. Ross.	... Alexander Hunter.
... Samuel Walker.	... James Rousseau.
... John Lamont.	1877 Thomas Adams.
... Robert Leitch.	... Thomas Fraser.
1875 Thomas J. Boyd.	... James Crichton.
... Hector M. Wilson.	... Wastell Arrowsmith.

HOUSE BURSARS.

(Continued from page 381 of History of Heriot's Hospital.)

1872 Henry Burnet.	1874 William Therburn.
... James F. Bannerman.	... Peter B. Gunn.
1873 George Rowley.	1875 Thomas M. Glen.
... Thomas B. Stuart.	1876 James B. Dunn.
... David Lyon.	... Robert Carmichael.
... David Callum.	... Thomas Taylor.
... Donald F. M'Donald.	... Robert S. M'Dougall.
... William Porteous.	... Thomas Porteous.
... David Syme.	... Clement Gunn.
... Peter B. Ritchie.	1877 John H. M'Donald.
... Hugh Falconer.	... George Chisholm.
1874 William M. M'Lean.	... Alfred Tod.
... Alexander Dewar	

ABERCROMBIE BURSARS.

(Continued from page 382 of History of Heriot's Hospital.)

1872 James A. Lyon, M.A.   1876 William A. Barclay.

HOSPITAL MEDALLISTS.

(Continued from page 400 of History of Heriot's Hospital.)

1872	James F. Bannerman.	1875	Clement B. Gunn.
...	Thomas R. Stuart.	...	Thomas Porteous.
1873	David B. C. Callum.	...	Robert Carmichael.
...	Hugh Falconer.	1876	Thomas Purves.
1874	Alexander Dewar.	...	John H. M'Donald.
...	William M. M'Lean.	1877	George R. Howison.
		...	Robert S. Thornton.

II.—CHRONOLOGICAL LIST OF THE MINISTERS WHO HAVE PREACHED THE COMMEMORATION SERMON ON THE ANNIVERSARY OF GEORGE HERIOT'S BIRTH-DAY.

(Continued from page 410 of History of Heriot's Hospital.)

1873 Thomas Gentles, M.A., Trinity College.
Hebrews xii. 16.

1874 John Barclay, Tron.
Psalm cxxxix. 14.

1875 Robert Horne Stevenson, D.D., St George's.
Matthew xi. 4.

1876 Alexander Williamson, West St Giles'.
Acts x. 38.

1877 Norman M'Leod, St Stephen's.
Galatians vi. 7.

III.—CHRONOLOGICAL ENUMERATION OF BOYS ADMITTED INTO HERIOT'S HOSPITAL SINCE 1872.

(Continued from page 370 of History of Heriot's Hospital.)

Year	Admitted	Year	Admitted	Year	Admitted
1872	32	1873	40	1874	28
1875	36	1876	25	1877	36

IV.—REVENUE AND EXPENDITURE OF HERIOT'S HOSPITAL.

(Continued from page 229 of History of Heriot's Hospital.)

REVENUE.		DISBURSEMENTS.	
1872	£22,436 16 8	1872	£11,792 4 2
1873	19,368 13 5	1873	12,773 3 7
1874	20,266 18 0	1874	12,688 6 5
1875	19,465 2 5	1875	12,762 10 9
1876	20,434 6 10	1876	12,388 0 0
[1] 1877		[1] 1877	

The following is an Analysis of the Hospital Expenditure during the year 1876, comprehended under the head of Ordinary Disbursements :—

Apprentices'. Fees,	£1211	0 6
Bursaries,	561	9 10
Clothing, Victualling, and Washing,	3184	18 7
Books and Stationery,	241	5 3
Coal and Gas,	234	0 9
Household Furnishings,	167	11 2
Repairs, Public Burdens, Feu-duties, Insurance, &c.,	1226	0 4
Salaries, Wages, and Annuities,	4796	11 0
Law Expenses, Incidents, Superannuation Scheme, &c.,	765	2 7
	£12,388	0 0

[1] The Accounts for 1877 had not been officially Audited when these Supplementary pages were going through the press. The Revenue for the year was upwards of £21,000, but it was impossible to submit any statement as regards the Disbursements until such had been approved of by the Accountant, and passed by the Auditors' and the Finance Committee.

c

The estimated Annual Expense for the *Maintenance* (including Victualling, Clothing, Washing, Coal, Gas, and Household Furnishings) and *Education* (including Books, Stationery, Salaries,[1] Wages and Annuities) of each boy during 1876 was £37, 16s. 9d.

SURPLUS REVENUE AVAILABLE FOR HERIOT SCHOOLS.

The following exhibits the surplus sums annually applicable for School purposes since 1871. The annual surpluses thus applicable from 1836 to 1871 will be found on pp. 278-9 of *History of Heriot's Hospital* :—

EXPENDITURE.

1872	.	.	.	£9684	5 11
1873	.	.	.	5795	9 10
1874	.	.	.	7528	8 9
1875	.	.	.	6699	1 8
1876	.	.	.	8009	10 10

---

[1] Under this head have been included the Salaries of all the officials except the Treasurer and his Clerks, the Clerk and Law-Agent, the Superintendent of Works and his assistants, and the Accountant; such proportion only of their salaries having been included as has been allocated by the Finance Committee to that portion of their duties which applies exclusively to the Hospital.

V.—TABLE OF HERIOT SCHOOLS.

Name of School.	Date of Opening.	Accommodation.	Principal Class Room.		Class Room.		Class Room.		Sewing Room.		Writing Room.	
			Length.	Breadth.	Length.	Breadth.	Length.	Breadth.	Length.	Breadth.	Length.	Breadth.
Heriot Bridge,	1838	318	58¾	31⅓	21	9½	...	...	31½	16½	...	...
Borthwick Close,	1840	320	60	28½	...	...	...	...	29¾	21	29¾	18¾
Old Assembly Close,	1840	320	60	28½	...	...	...	...	29¾	21	29¼	18¾
Cowgate Port,	1840	330	54	29	9⅓	10	...	...	29	20	19⅓	16½
High School Yards,	1840	330	54	29	9⅓	10½	...	...	29	20	19⅓	16½
Brown Square,	1846	240	47	24½	24	16½	...	...	24	16½	...	...
Rose Street,	1848	326	42	37½	22½	20	...	...	28½	21	...	...
Broughton Street,	1855	250	34½	22¼	21¾	12	...	...	26½	19½	26½	18¼
Grindlay Street,	*	290	...	...	...	...	...	...	...	...	...	...
Abbeyhill,	1875	320	55	23	21½	18	...	...	20	18	20	17
Davie Street,	1876	400	51	23	20	18	16½	13	20½	21½	21½	21
Stockbridge,	1877	400	55½	23½	20½	18	16	13½	23	20	23	20
INFANT SCHOOLS.												
High School Yards,	1840	185	30	27½	17¼	11¼	...	...	...	...	...	...
Rose Street,	1848	170	43	27	...	...	...	...	...	...	...	...
Broughton Street,	1855	160	34½	21	21	12	...	...	...	...	...	...
Victoria Street,	1866	330†	46½	27⅓	46½	27⅓	...	...	...	...	...	...
Abbeyhill,	1875	180	50½	25	...	...	...	...	...	...	...	...
Davie Street,	1876	200	51	23	21	20	...	...	...	...	...	...
Stockbridge,	1877	200	55½	23½	20½	18	...	...	...	...	...	...
		5329										

* This is a temporary School, established for the convenience of the Fountainbridge district.
† This number includes 150 in a Supplementary Class.

VI.—LIST OF THE GOVERNORS AND OFFICIALS OF HERIOT'S HOSPITAL, FEBRUARY 1878.

MAGISTRATES AND TOWN-COUNCIL OF EDINBURGH.

Lord Provost—Right Honourable THOMAS JAMIESON BOYD.

Bailies.

THOMAS ROWATT.
JAMES COLSTON.
JOHN TAWSE.

ROBERT CRANSTON.
WILLIAM ANDERSON.
GEORGE ROBERTS.

Dean of Guild, ... JOHN SMITH.
Treasurer, ... JOHN WILSON.
Convener of the Trades, . ROBERT LEGGET.

Councillors.

THOMAS DRYBROUGH.
JAMES CRIGHTON.
JOHN HOPE.
ROBERT WHITE.
THOMAS HALL.
WILLIAM GILMOUR.
JOHN CLAPPERTON.
JAMES GOWANS.
JAMES STEEL.
GEORGE HARRISON.
DANIEL SUTHERLAND.
JAMES MACKNIGHT.
ALEXANDER HENRY.
ARCHIBALD SUTTER.
JOHN BOYD.

ADAM MOSSMAN.
ROBERT YOUNGER.
JAMES BUCHANAN.
ROBERT MITCHELL.
THOMAS LANDALE.
JOHN MILNE.
ROBERT HAY.
ROBERT SOMERVILLE.
THOMAS CLARK.
ROBERT MACDOUGALD.
JOHN BAXTER.
WILLIAM B. M'LAUCHLAN.
ROBERT GORDON.
ROBERT TURNBULL.
ALEXANDER REID.

MINISTERS OF EDINBURGH.

WILLIAM ROBERTSON, D.D.
WILLIAM H. GRAY, D.D.
JOHN STUART, D.D.
ROBERT H. STEVENSON, D.D.
ARCHIBALD SCOTT, D.D.
CORNELIUS GIFFEN.
THOMAS GENTLES, M.A.

ALEXANDER WILLIAMSON.
W. C. E. JAMIESON, B.A.
NORMAN MACLEOD.
JOHN GLASSE, M.A.
JOHN WEBSTER, M.A.
JAMES C. LEES, D.D.

OFFICIALS.

Treasurer, ...	DAVID LEWIS.
Clerk, ...	GEORGE BAYLEY, W.S.
Superintendent of Works,	JOHN CHESSER.
Accountant, ...	JAMES M. MACANDREW.
Treasurer's Office—	
Cashier, ...	GEORGE LYON.
First Clerk, . . .	GEORGE A. BARCLAY.
Second Clerk, . .	GEORGE A. LAMB.
House-Governor, . . .	FRED. W. BEDFORD, LL.D., D.C.L.
Physician, ...	Sir ROBERT CHRISTISON, Bart., M.D.
Surgeon, ...	ANDREW WOOD, M.D.
Dentist, ...	WILLIAM A. ROBERTS, M.D.
Apothecary, ...	JAMES GARDNER.
Matron, ...	Miss SARAH J. HERRON.

MASTERS.

Classics, ...	JOHN RIDPATH, M.A.
Arithmetic & Mathematics,	HENRY G. C. SMITH.
English, ...	{ WILLIAM G. WILSON, M.A. JOHN CRAIG, M.A. DONALD FERGUSON.
Writing, ...	JAMES WATSON.
French, ...	JULES A. L. KUNZ.
Drawing, . . , . .	JAMES B. NAPIER.
Vocal Music, ...	T. M. HUNTER.
Dancing and Calisthenics,	GEORGE LOWE.
Shorthand, ...	JOHN THOMPSON.

Steward, ...	JOHN ROBERTSON.
Gatekeeper, ...	JOHN LEVICK.
Gardener, ...	ALEXANDER M'DONALD.
House Carpenter, . . .	PETER DOCHERTY.
Wardsmen, ...	{ ROBERT EMSLIE. GEORGE HUTTON. JAMES GRAHAM.

VII.—LIST OF THE HERIOT FOUNDATION SCHOOL TEACHERS in FEBRUARY 1878.

Inspector—FRED. W. BEDFORD, LL.D.

JUVENILE.

Name of School.	Head-Teacher.	Sewing-Mistress.
Heriot Bridge, .	WILLIAM ROBERTSON.	Miss H. H. SUTHERLAND.
Borthwick Close, .	GEORGE G. BARKER.	Miss GRACE DICK.
Old Assembly Close,	ROBERT MAUCHLINE.	Miss ELIZABETH GARDINER.
Cowgate Port, . .	JOHN BYERS.	Miss NAOMI ORMISTON.
High School Yards,	JAMES M'KEAN.	Miss ISABELLA AITCHISON.
Brown Square, .	JOHN DUNNETT.	Miss ANN DUNCAN.
Rose Street, . .	WILLIAM WINSTANLEY.	Miss JESSIE URQUHART.
Broughton Street,	THOMAS ARMSTRONG.	Miss ADELAIDE Z. SMITH.
Grindlay Street, .	ANDREW KERR.	Miss AGNES CARFRAE.
Abbeyhill, . . .	JOHN MARSHALL.	Miss ISABELLA GARDEN.
Davie Street, . .	JOHN M'CRINDLE.	Miss CATHERINE MUNRO.
Stockbridge, . .	WILLIAM HALL.	Miss LAVINIA LAING.

INFANT.

High School Yards, . . .	Miss JANET MILLAR.
Rose Street, ...	Miss ISABELLA ANDREW.
Broughton Street, . . .	Miss ANN BELL.
Victoria Street, . . .	Miss HELEN H. THOMSON.
Abbeyhill, ...	Miss CHARLOTTE GARDEN.
Davie Street, ...	Miss JANE JOHNSTON.
Stockbridge, ...	Miss JANE WIGHTON.

MUSIC TEACHERS.

Mr THOS. SMITH.	Mr T. M. HUNTER.	Mr JOHN HALL.
Mr J. C. GRANT.	Mr W. KERR.	Mr JAMES SHARP.

VIII.—COLLECTIONS, AND NOTES, HISTORICAL AND GENEALOGICAL, REGARDING THE HERIOTS OF TRABROUN.

INTRODUCTION.

The derivation of the word " Heriot " is given in Jamieson's Scottish Dictionary (Longmuir's edition) thus :—" HERIOT : The fine exacted by a superior on the death of his tenant. From Anglo-Sanon *heregeat*, compounded of *here* exercitus and *geat-an*, reddere, erogare. This primarily signified the tribute given to the lord of the manor for his better preparation for war; but came at length to denote the best *aucht*, or beast of whatever kind, which a tenant died possessed of, due to his superior after death. It is therefore the same with the English forensic term *Heriot*. Here we have the meaning of the surname of George Heriot."

There is a parish, and a river in it, both of the name of Heriot, about twenty miles south-east of Edinburgh ; also a small stream of the name in the parish of Cockburnspath, Berwickshire ; but how the name came to be applied to any of them does not appear to be ascertained. Chalmers, in his " Caledonia," says—" The origin of the singular name of this parish is uncertain. Heriot, probably, is neither the original name of the water (river), nor a descriptive appellation of the place. . . . Heriet is the spelling in the ancient *Taxatio ;*" and he states his opinion to be that the name is derived from *Hergeath* in the Anglo-Saxon, signifying " an invasion—a spoliation."

There is likewise " a very old artificial mound or embankment called Herit's Dyke [1] mentioned by Chalmers, and also in the statistical accounts of the parishes. It is supposed to have been erected by the Romanised *Ottadini*. Perhaps this dyke was not intended to be a military work ; it may only have been a boundary fence separating one large district of country from another. This embankment, Chalmers says, had been, shortly

---

[1] The Dyke is called *Herriot's* in the Statistical Account of the Parish of Greenlaw, and *Harit's* in that of Westruther. It may be noticed that Sir Walter Scott, in a note to his " Provincial Antiquities" (Edition 1853, p. 253) says—" The learned William Hamper, of Birmingham, has sufficiently proved that the word *Hare* or *Har* refers to a boundary."

before he wrote [1810] traced for fourteen miles, running in a south-east direction through the parishes of Greenlaw and Westruther."—D. Milne Home, F.S.A., in P. of S. of Antiq., 1870–72, vol. ix. p. 469.

The first notice believed to have been yet found of the name (or what is very likely the name), used as a surname, is that of "Willo de Heryt." who was a witness to a Charter granted by King William the Lion, prior to 1214,[1] and probably the surname was assumed from some connection with the parish or dyke above mentioned.[2]

In Wynton's "Orygnale Chronykil" of Scotland, completed about 1426, he states that John Gibson, "that wes gud man, and William Heryot, that wes then [1334] duelland in till the Barony" of Rothsay, isle of Bute, effected the escape of Robert, High Steward of Scotland, from Rothsay, whither he had fled from the English after the battle of Halidon Hill. The escape of the Steward, afterwards Robert II., proved to be one of the most important events in Scottish history. After his escape he rallied the Scots, and Tytler, the historian, says "He was the main instrument in defeating the designs of David the Second and Edward the Third when an English Prince was attempted to be imposed upon the nation."

By Charter, dated in 1423, Archibald Earl of Douglas conveyed the lands of Trabroun to John Heriot, son of James Heriot of Niddry-Marshall. This latter place is a few miles east of Edinburgh, and is said to have received its appellation "from the Wauchopes, who in ancient times were guardians of part of the south borders of Scotland, and hence were denominated Marshalls. By this means it was distinguished from Niddry in East Lothian, called Longniddry, and from Niddry-Seaton in West Lothian, which two centuries ago was the property of the Seatons. The Heriots were once proprietors of Niddry-Marshall, at least a part of it, for they had the title of Niddry-Marshall assigned to them." [3]

There does not appear to be any evidence extant of a relationship between William Heriot, mentioned by Wynton, and James Heriot of Niddry-Marshall, but as a title and a grant of land were rewards frequently bestowed in ancient, as well as modern, times for distinguished services, it is not improbable that William Heriot obtained the title of Marshall and part of

---

[1] "Liber de Melros."—Ban. Club, vol. i. p. 49–50.

[2] Professor Innes, in his work entitled "Concerning some Scotch Surnames," classes the surname Heriot among those derived from places or lands.

[3] Trans. of Soc. of Antiq., vol. i., 1792.

the lands of Niddry for his aid in the escape of the High
Steward from the English ; and thus may have been one of the
ancestors of James Heriot of Niddry-Marshall and of the
Heriots of Trabroun.

LIST OF ABBREVIATIONS.

A. D. A., ...	Acta Dominorum Auditorum.
A. D. C., ...	Acta Dominorum Concilii.
P. C. T., ...	Pitcairn's Criminal Trials.
R. of D., ...	Register of Deeds.
R. of P. C., ...	Register of Privy Council.
R. of P. S., ...	Register of Privy Seal.
R. of R.	Register of Retours.

---

te 1423.   I. JAMES HERIOT of Niddry-Marshall.

1432.   II. JOHN HERIOT of Trabroun (his son).

Obtained a grant by Charter, dated 2d December 1423–4, of the
lands of Trabroun, near Lauder, Berwickshire, from Archibald
Earl of Douglas, which Charter was confirmed by James I., 8th
January 1423–4. The King's Charter styles him, " *dilecto armi-
gero suo Johanni de Heriot, filio ac heredi dilecti confederati sui
Jacobi de Heriot de Nidri-Marshall.*"

rca 1440   III. SYMON HERIOT of Trabroun.

Douglas' Baronage.

*ir ca*1460   IV. JAMES HERIOT of Trabroun (his son).

Married in the year   the elder daughter of Patrick
Congalton, younger of Congalton, near Haddington.—*Ibid.*

1480. Was one of the judges in an inquest held at Edinburgh before
Sir Patrick Hepburn, Governor of Berwick, on claim by the
Abbot and Convent of Melrose for houses in the Briggate of
Berwick. His opinion was " that they should have a house at
the corner to the Tweed."—*Liber de Melrose, Ban. Club,* vol. ii.
App.

1480-83. Died between 1480 and 1483.

1483. The Lords Auditors decreed William, son and apparent heir of Gabriel Towris, to pay to Robert Heriot, executor to umquhile James Heriot, ten merks, taken by him from the said James and Robert for the mails of the lands of Muirhouse, and two horses, price five pounds, an ox, price 35 shillings, all taken by the said William out of the said lands, with six shillings costs of three witnesses, and twenty shillings[1] costs of petition.—*A.D.A.*

V. ANDREW HERIOT of Trabroun.

(Son of the preceding James Heriot of Trabroun.)

1488. Raised an action before the Lords of Council against George Lord Seton for 20 merks paid to him to put him (Andrew) in fee (as successor) in the lands of "Auldinstoun." After parties and their procurators had been heard, Andrew agreed that if he got his grandfather's consent he would allow the said George to retain the said 20 merks provided he was put in fee of the said lands, which was agreed. And the Lords decerned accordingly, under penalty in case of failure.—*A.D.C.*

1515. Summoned along with Lord Hume and his four brothers to appear in October to hear themselves "forfawt for causes."—*P.C.T.*

1515. Had forfeited his lands of Trabroun for treason, and was restored to them and all his privileges by Act of Parliament of this date (1515).

1527. Summoned by "our Souerane Lord," and John Legate, grandson and heir of Andrew Legate, for the hasty execution of his office of Sheriff-Depute (Principal) of Edinburgh, within the Constabulary of Haddington, in giving sasine of a saltpan and house asked for by William Sinclair of Herdmanston, which were alleged not to belong to him. At same time, James Heriot, George Heriot, the said William Sinclair, and others,

---

[1] All *Scots* money.

were summoned, because in an inquest they found that the late John Sinclair of Herdmanston, great-grandfather of the said William, died last "vestit and seisit" in the saltpan and house, while it was alleged that, long after his decease, the late Andrew Legate, grandfather of the said John, died "vestit and seizit" of them.

The Lords of Council acquitted the said William Sinclair and persons of inquest because the sasine given by the Sheriff-Depute was conform to a precept of sasine shown to the Lords of date 9th November 1525.—*A.D.C.*, v. 37.·

1527. Respited with fourteen others for treasonably arraying against the King beside Linlithgow.[2]—*P.C.T.*

1527. Granted, along with his wife, "Marione Cokburne," a renunciation of eight acres of the lands of Lethington, near Haddington.—*A. of P. in favour of Earl of Lauderdale.*

1529. His spouse, Marion Cockburn, and he, for his interest, had "Letters Purchased" (Fr., *pourchasser*, to pursue) against them by Patrick Hepburn, Master of Hales, Sheriff-Depute of Edinburgh, because she rebelled against him in poinding her goods, for not having given suit and presence in the Sheriff Courts, on the ground that she was discharged therefrom ; and also to make her produce the Letters Purchased by her to that effect.

The Lords of Council deemed her Letters to be orderly and just, proceeding on a Royal discharge from giving suit and attendance in Sheriff Courts ; and ordained the Sheriff-Depute to cease from executing his office against her.—*A.D.C.*, v. 40.

1529. Was infeft in half of the lands of Michelston.   Witnesses, David Heriot, Mr Thomas Heriot, &c.—*Protocol Bk. at Hadn.*

---

[1] Probably the "treasonably arraying" was being engaged in the following attempt :—

1525—James V.—"A feeble attempt was indeed made by Arran to prevent by force the ratification of the truce ; and, for a moment, the appearance of a body of 5000 men, which advanced to Linlithgow, threatened to plunge the country into war ; but the storm was dissipated by the promptitude of Douglas. Taking the King along with him, and supported by the terror of the royal name, he instantly marched against the rebels, who, without attempting to oppose him, precipitately retreated and dispersed."—*Tytler's Hist.*

1529. Convicted, along with Gilbert Wauchope of Niddry-Marshall and others, of having been "art and part" (or aider and abettor) of convocation of the lieges upon Edmonston of Edmonston.[1] Sentence not recorded.—*P.C.T.*

He was twice married—(1) to Janet Borthwick; (2) to Marion Cockburn; and died in 1530 or 1531.

In the Index of Testaments of the Commissariat of Edinburgh there is the subjoined entry, but, unfortunately, the record itself corresponding to this period is not now extant :—

"Honorabilis vir Andreas Heriot. Dominus de Trabron apud Heprig, 2 July 1532."

JAMES HERIOT, "of the family of Trabroun," and, probably, son and apparent heir of Andrew Heriot of Trabroun, whom he predeceased.[2]

Was uncle to George Buchanan, the celebrated poet and historian, and sent him, in or about the year 1520, to the University of Paris to complete his studies. After he was there two years his uncle died, and he returned to Scotland poor and in bad health. His own words (in his *Vita ab ipso*) are, "*intra biennium avunculo mortuo et ipse gravi morbo correptus ac undique inopia circumventus ad suos est coactus.*"

AGNES HERIOT, "of the family of Trabroun," and sister of the preceding James Heriot.

circa 1500 Was married to Thomas Buchanan of Moss, Stirlingshire. Their third son, born in February 1506, was the celebrated

---

[1] At the time there was a deadly feud between the lairds, Edmonston of . Edmonston and Wauchope of Niddry-Marshall.

[2] In the beginning of the 16th century, and till at least 1520, there was a James Heriot "Justiciar" of Lothian. He is styled "*Canonicus Rossensis ac officialis Sancti Andree infra Archidiaconatum Laudonie judex,*" and in 1518 was judge on a claim by the chaplain of the parish of Crichton against Andrew Heriot of Trabroun (his father?) for a competent *mortuarius* for his deceased wife Janet Borthwick.—*Liber Officialis Sancti Andree*, Abb. Club.

Dr Irving, in his Memoir of Buchanan, says his uncle James Heriot sent him, "apparently in the year 1520," to Paris, and died two years afterwards (in 1522), and as that is the year in which another judge, "William Prestoun, Rector de Beltoun," appears on record in place of Heriot, there is a presumption that James Heriot the Justiciar, and James Heriot uncle of Buchanan, were the same person.

George Buchanan already mentioned. A place in the parish of Killearn, " which had been adapted to the purpose of shielding her flock, is still denominated Heriot's Shiels."—*Irving's Mem. of G. B.*

VI. JAMES HERIOT of Trabroun.

(Grandson 'and heir of Andrew Heriot of Trabroun, who died in 1530 or 1531.)

1531. Obtained by letter from King James V. under the Privy Seal (in which he is styled grandson and heir of umquhile Andrew Heriot of Trabroun) the gift of the ward " of All and hale the landis of Trabroun, with the pertinentis liand within the Sheriffdome of Berwick now being in our Souerane lordis hands be ressoun of ward be deces of the said umquhile Andro."

1531. Was infeft in husband land (about 26 acres) and oxgang land in Langniddrie. John Cockburn of Ormiston acted as Bailie. Witnesses, James Heriot, George Heriot, David Heriot in Haddington.

Same day and hour, Brother John [Cockburn] made oath that he would not hurt Marion Cockburn, spouse of Andrew Heriot of Trabroun, in her terce of said lands.—*Protocol Bk. at Hadn.*

1535. Acquired from Marion Cockburn, relict of Andrew Heriot of Trabroun, her eight merkland of her husband land in Langniddrie.—*Ibid.*

1537. Renounced an annual rent of eight merks out of the lands of Leithington.—*A. of P.*, 1661.

1542. Was infeft in the principal mains of the middle third of the lands of Audneston, in the barony of Tranent, on precept of sasine by George Lord Seytoun.—*Hadn. Burgh Record.*

1549. Pardoned for having been one of the accomplices of the Earl of Glencairn in his treasonable attempt against the Lord Governor

on the moor of Glasgow, in the month of May 1544.[1]—
R. of P. S.

1550. Was, with certain sheriffs and lairds, required to assist in furnishing oxen and pioneers for the bringing of munition and
artillery to the " oist " and army ordered to assemble at Edinburgh, 16th April (1550).—*R. of P. C.*

1551-2. Obtained a precept for a charter, to be expede under the Great
Seal, confirming charter by the then deceased Patrick Cranston,
of " Ratho Byris," in favour of his (J. H.'s) grandfather and
grandmother, in which he is styled " beloved servant of our
Lady the Queen," to the lands of Arros, lands in Harlaw, &c.,
lying in the regality of Lauderdale and sheriffdom of Berwick.
—*R of P. S.*

1553. Obtained from the Crown, along with John Hamilton, a gift of
the escheat goods which pertained to Robert Reidpeth in
Nethershole.—*R. of P. S.*

1553. Obtained a precept for expeding a charter under the Great Seal,
confirming charter from John Tennant of Listoun Schelis, of the
lands of Over Howden, lying within the lordship of Lauder and
sheriffdom of Berwick.—*R. of P. S.*

1554. Pardoned, along with others, for treason.—*Ibid.*

1554. Obtained an obligation (dated 21st November) from Alexander
Burnet, of Leys, for the sum of nine score merks due to the
" Lord Sanctandrois " (Archbishop) for the feu (perpetual
annualrent) of the lands of Invery and half of Kirkton of
Banquhar. Witnesses, Alexander Heriot and others.—*R. of D.*

1554-5. As curator for Thomas Hamilton, son and heir of deceased
Thomas Hamilton, of Priestfield, let by deed, dated 1st Febru-

---

[1] 1544-5. Feb. 6 & 14. Sir Andrew Heriot, chaplain in Glasgow, found
caution to underly the law for the same offence.—*P. C. T.*

ary, to Peter Dundas, the two parts of the lands of Balvin, within the sheriffdom of Perth, for the space of three years, for certain quantities of grain, or the value thereof, yearly.— *R. of D.*

1555. Was a consenter to a Contract, dated 8th July, between John Couttis, Burgess of Edinburgh, tutor to John Lawson of Lochtullo, and Alexander Home, son to William Home of Prendergaist, concerning the lands of Denisleeside, Lochtullo. Witnesses—Alexander Heriot, and George Home, yr. of Spott.— *R. of D.*

1556. Infeft in lands of Frierness in Lauderdale on Deed by Alexander Lord Hume. Witnesses—George Heriot, in Longniddrie, Symon Fortoun, Jacobus Heriot de Hirnyclewt, &c.—*Protocol Bk. of Hadn.*

1556-7. Was surety along with Gilbert Wauchope of Niddry-Marshall and Patrick Hepburn of Wauchton that William Wauchope (son and heir-apparent to Gilbert), Edmund Nicholson, miller in Dirleton, and five others, would underly the law for slaying wild fowl with culveringis and pistolettis, between May 1552 and November 1556.—*P.C.T.*

1557. As one of the Curators of Thomas Hamilton of Priestfield (son and heir of the deceased Thomas Hamilton of Priestfield), was party to a Contract, dated   1557, between Elizabeth relict of the deceased Thomas Hamilton and William Hutsoun her spouse.—*R. of D., v. 3. f. 6.*

---

[1] Nisbet in his *Heraldry* (originally published in 1722), speaking of the Heriots of Trabroun, says, "of whom [them] are the Heriots in Longniddry," and this statement is corroborated by the circumstance of George Heriot in Longniddry having been a witness (as will be seen in these pages) to important deeds connected with the family. The representatives of the Heriots, farmers, at Castlemains, Dirleton, about the end of last century, some of whom are in Gullane, North Berwick, and South Carolina, U.S., claim descent from this George Heriot. It may also be noted that George Heriot in Longniddry obtained a lease under the Privy Seal, dated 14th March 1553-4, of the East and West Barnes Links (the sandy ground at the sea-shore near Dunbar.)

1559. Granted a discharge on behalf Lord Sanct Androis (the Arch-bishop), to Master[1] William Roy, for receipt of debt due by the Earl of Huntly. Deed dated 1st June. Witnesses—Thomas Fawsyde of that ilk, George Heriot in Longniddry, Alexander Heriot in Laurieston, and others.—*R. of D., v.* 3, *f.* 240 *B.*

1560. Was one of the Commissioners for the Burghs in the memorable Parliament of Queen Mary, held at Edinburgh, 1st August.

1560. Entered into a Contract, dated 20th August, with Master John Scrymgeour[2] of the Myres, and William Scrymgeour, his son, concerning debts, etc.—*R. of D., v.* 8, *f.* 401 *B.*

1560-1. Granted a Discharge, dated 8th March 1560, to Sir Richard Maitland of Lethington, Knight,[3] for receipt of money due by him per contract of marriage between Isobel Maitland, his daughter, and James Heriot, younger of Trabroun. Witnesses —Patrick Cockburn of Clerkington, John Maitland, and others. *R. of D., v.* 2, *f.* 89.

1562. Was one of the prosecutors in the trial of William Ferguson and William Wright, Restalrig, convicted of aiding and abetting the cruel and unmerciful slaughter of John Borthwick in Restalrig.[4]—*P.C.T.*

---

[1] Master or Mr at this period signified the degree of M.A.

[2] Mr John Scrymgeour was Laird of Myres, near Auchtermuchty, Fife-shire, and held the office of Master of Works to King James V., with whom he was in great favor. He designed and built the Palace of Falkland, which is still in good preservation, and may long remain a monument of his archi-tectural skill and taste.

[3] Father of the famous William Maitland, Secretary to Mary Queen of Scots, and an ancestor of the present Earl of Lauderdale.

[4] The murderers were found guilty, and beheaded on the Castle Hill of Edinburgh.

"This is the first instance where the Editor has, in the earlier Records of this Court, met with the mention of a sword having been employed by the public executioner by order of the Justiciar for decapitation. It is by no means improbable that death had been inflicted on the murdered man by that instrument, and that the Judge, to make the punishment still more striking, had judicially ordered the selfsame weapon to be used for their decapitation." —*Note by Editor of P.C.T.*

1564. Was one of the prolocutors (Advocates) for Alexander Haitlie and others, acquitted of the slaughter of Steven Brounfield, younger of Greenlawden.—*P.C.T.*

1565. Was one of the prolocutors for William Sinclair of Herdmanstoun and others, accused of the slaughter of Walter Murray, servant to James Earl of Bothwell.   Verdict—" Acquit the baill."—*P.C.T.*

1566. Was cautioner for James Earl of Bothwell, in bond granted by him for payment of 500 merks yearly to Sir Simon Preston of that ilk, and Simon his son, " attour ye ordinar appointit to him (Sir Simon) be oure Souerane for ye keeping of ye Castle of Dunbar."—*Orig. in Reg. Ho.*

1567. Was surety that William Douglas of Cavers would enter himself within twenty-four hours in ward within the Castle of Blackness, there to remain upon his own expenses until liberated by the Lord Regent, under the penalty of 2000 merks. —*R. of P. C.*

1567. Was surety that Thomas Turnbull of Hassendeanbank would remain in ward within the Burgh of Edinburgh on his own expenses until relieved, under the penalty of 500 merks.— *R. of P. C.*

1568. Excused for non-attendance in Parliament of 8th August, as Mr Clement Little alleged he was in ward at command of my Lord Regent Governor, in keeping of the Lord Home, within the Castle of Home and Fast-castle, and " sua mycht noct compeir." —*A. of P.*

1568. Was taken prisoner by the Regent's forces at the Battle of Langside.—*Tytler's H. of S.*

1571. Was one of the prolocutors for Lewis Lumsden Dysart, and John Lumsden his son, accused of the slaughter of David Erl,

d

tailor in Dysart, committed at the Kirk of Dysart. Verdict—
" The Assyse acquit the pannell."—*P.C.T.*

1571-2. Made faith, and Thomas Fawsyde of Fawsyde became security
for his entry (in ward ?), and the rendering (giving up) of his
house if charged to do so, under the penalty of £1000 Scots.—
R. of P. C.

1574. Was one of the six Commissioners appointed by the General
Assembly of the Church of Scotland to wait upon the Lord
Regent, and present to his Grace the heads and articles which
the Assembly had put in writing.—*Calderwood's Hist.*

1577. Was one of the prolocutors for John Crawford of the Schaw,
and others, accused of fire raising, and the burning of " ane
byir," belonging to John Boswell of Auchinleck, and · other
crimes. The trial seems to have been abandoned, as the Laird
of Auchinleck " compeirand disasentit to the pursute of thame."
—*P.C.T.*

Married Janet Cockburn, of the ancient family of Cockburn
of Ormiston.

1580. Died on 4th October.
His Will is dated 11th August same year (1580), and recorded 27th June 1581, with an " Eik " by his executor on 12th
December 1583. The residue of his personal effects, which consisted chiefly of agricultural stock, and debts due to him,
amounted to £6373, 14s. 1d. Scots. His heritable (or real)
estate may have been disposed of by deeds different from those ·
which conveyed his personal property.

JANET HERIOT (Grand-daughter of Andrew Heriot of
Trabroun).

1542. Discharged her brother James, of Trabroun, of certain sums of
money contained in, and left to her by, her grandfather's and

father's testaments.   Deed signed at Elvingston.[1]   Witnesses,
Sande Heriot, David Heriot, &c.—*Protocol Bk. at Hadn.*

1553. Married John Acheson,[2] a captain in the Scots Guards to the
King of France, wrote in 1553 to the Queen Regent of Scotland in his behalf, and referred to the circumstance of his being heritable proprietor of certain lands near the Mylnhaven
(now called Morisons-haven), between Musselburgh and Pres-
tonpans.

Was described in a sasine as a widow, residing at Prestoun in
1560, liferented in a house and certain lands, which had been
granted in heritable fee to her husband's father.—*Communicated.*

ADAM HERIOT, "of the Family of Trabroun."

1559. "Adam Heriot, of the family of Trabroun,[3] in East Lothian,
born about 1514, was a conventual brother of the Augustinian
Order of the Abbey of St Andrews, who embraced the Protes-
tant faith.   He was removed to Aberdeen in 1560,[4] yet retained
the vicarage of St Andrews till his death in 1574."—*Scott's Fasti
Ecclesiæ Scoticanæ.*

1561. Named among the preachers of the Reformation whom Quintin
Kennedy, Abbot of Crossraguel, in " Ane Oration" denounces
as " pestilent precheouris puffit up with vane glore," &c.

---

[1] The family afterwards called part of the estate of Elvingston, Trabroun,
and built a mansion upon it, which was taken down about forty years ago.
The property at present belongs to Robert Ainslie Esq.

[2] An ancestor of the present Earl of Gosford, Ireland.

[3] The precise relationship has not been ascertained.   But as he was an
ecclesiastic in St Andrews, and as James Heriot (in 1520) was an official of St
Andrews and judge within the Archdeaconery of Lothian, and James Heriot of
Trabroun acted (at least on two occasions—1554 and 1559) as a factor or col-
lector for the Archbishop, the statement that he was of the family is likely
to be correct.

[4] "1560.   On 16th July the French embarked, and the same day did the
English army depart towards Berwick.   The third day after their parting a
solemn thanksgiving was kept in the Church of St Giles, Edinburgh, by the
Lords and others professing true religion, and then were the ministers by common advice distributed among the Burghs.   JOHN KNOX was appointed to
serve at Edinburgh, CHRISTOPHER GOODMAN at St Andrews, ADAM HERIOT
at Aberdeen," &c.—*Spotswood's. Hist.*

The Abbot likewise in his "Compendious Ressonyng" says :—"Sextlie, I will desyr Hereot (*qui adhuc hesitat*) to mak ane confutatioun to oure confirmatioun groundyt upon the testimoniis of the New Testament to preif the figuris of the Auld Testament sufficient pruyf of materis of fayth concerning the New Testament."—*Knox's Works, Laing's Edn., v.* vi.

1566-7-8. Was a member of the General Assemblies of the Church of Scotland June 1566, July 1567, February 1568 ; and was one of the committee on the case of Paul Methven.

1569. Presented by the Crown with the parsonage and vicarage of the Kirk of Rachen (Rathen), Aberdeenshire.—*R. of P. S.*

1570. Appointed by the General Assembly "one of those to deal with the Earl of Huntly regarding the restoration of the collectors of the Kirk to their situations."

1574. Granted by the Crown a pension of £50 yearly during life. The warrant is dated at Holyrood House 7th April, and in it Heriot is styled besides "minister of Aberdeen, one of the Chapter of the Abbey of St Andrews."—*R. of P. S.*

BIOGRAPHICAL NOTICE.

" In the same month (August 1574) Adam Heriot, minister at Aberdeen, departed this life ; a man worthy to be remembered. He was sometime a friar of the Order of St Augustine, and lived in the Abbey of St Andrews, an eloquent preacher, and well seen in scholastic divinity. The Queen Regent coming on a time to the city, and hearing him preach, was taken with such an opinion of his learning and integrity, that in a reasoning with some noblemen upon the Article of the Real Presence, she made offer to stand to Heriot's determination. Warning of this being given, and he required to deliver his mind upon that subject in a sermon, which the Queen intended to hear, he did so prevaricate as all that were present did offend and depart un-

satisfied. Being sharply rebuked for this by some that loved
him, he fell in a great trouble of mind, and found no rest till he
did openly renounce Popery and join himself with those of the
congregation. Afterwards, when order was taken for distribution of ministers among the burghs, he was nominated for the
city of Aberdeen (in which there lived divers addicted to the
Roman profession), as one that was learned in scholastic divinity, and for his moderation apt to reclaim men from their
errors. Neither did he fail the hope conceived of him, for by
his diligence in teaching, both in the schools and church, he did
gain all that people to the profession of the truth. Fourteen
years he laboured among them, and in the end was forced by
sickness to quit his charge. He died of apoplexy 28th August
1574, in the sixtieth year of his age, greatly beloved of the citizens for his humane and courteous conversation, and of the
poorer sort much lamented, to whom he was in his life very
beneficial."—*Spotswood's Hist.*

VII. JAMES HERIOT of Trabroun (Son and Heir of last-
mentioned James Heriot of Trabroun).

1560. Married Isabel Maitland, daughter of the eminent Sir Richard
Maitland (formerly mentioned). Contract of Marriage dated 1st
October 1560. By this contract, and subsequent deeds, dated
21st February 1560-1, 27th September 1580, and 13th March
1583, the lands of Arrois, in Berwickshire, Elvingstoun, within
the constabulary of Haddington, Husbandland and one oxgate
of land in Longniddrie, were conveyed in liferent to the said
Isabel Maitland. These deeds were confirmed by charter under
the Great Seal 20th January 1586-7.—v. 36, p. 261.

1571. Apprehended when about to sail for France. The following
account of the apprehension is taken from *Bannatyne's Jour-
nal* :—

"Tuysday, the 18 of Septr. 1571.—George Auchinleck came
to Kinghorne and went aburde on William Sibbatis (Sibbald's)
shipp whar thair was the young laird of Trabroune and ane

vther called Borthick, sone to Michael Borthick, that is forfaltit, who were bound to France, but he tuik thame and thair writings also. The said Sibbat suld have been puneist for fals hard heides."

1580. Served (adjudged) heir to James Heriot, his father, in lands ot Henschawsyde, and Husbandland and one oxgate of land in Longniddry.—*R. of R.*

1580. Entered into a contract regarding certain lands with George and Peter Heriot, his brothers, executors appointed under the will of their father James, and with George, as apparent heir of his late brother Alexander Heriot. Contract dated 26th November. Witnesses, Thomas Buchanan, keeper of the Privy Seal ; George Heriot, goldsmith, burgess of Edinburgh ;[1] Andrew Keland, servitor to the Laird of Lethington ; David Lawtie, and Adam Lawtie, his son, and others.—*R. of D., v.* 19.

1583. Was guaranteed by Robert Fawsyde, younger of Fawsyde, that Sir William Lauder of Hatton, Knight, Gilbert Lauder of Balbouties, Thomas and John Lauder, his (Gilbert's) sons, William Lauder in Kelso, his brother; David Sinclair of Blanse ; and John Hoppringillis of Munyis, would be free and skaithless from him in their persons, families, dependents, and properties, under a penalty of 3000 merks.—*R. of P. C.*

1585. Included in the pardon granted by Act of Parliament to John Earl of Morton and his friends for all acts of hostility, "the murder of His Majesty's" "dearest father allanerlie exceptit."

1590. Was one of the Prolocutors for the prosecutors of James Tweedy of Drummelzeare and others, accused of having been aiders and abetters of the slaughter of Patrick Veitch, son of William Veitch, Dawick. Trial adjourned to the Justice Court of Peebles.—*P. C. T.*

---

[1] Father of the Founder of Heriot's Hospital.

1607. Served heir to James Heriot, his father, in the lands of Trabroun.—*R. of R.*

1611. Sold under reversion (with consent ot his wife Isabella Maitland) the lands of Trabroun to John Hamilton, his grandnephew, and son of Thomas first Earl of Haddington. The sale, as afterwards stated, was allowed to become absolute.

1618. June 4, Died. His Will is dated 22d October 1612, and recorded 3d April 1619. In it he states he is of "great aige," and leaves his soul to the eternal God, and his body to be honorably buried in the burial place of his father "at Hadingtoun Kirk, at the southe eist syd thairof."

His personal effects consisted chiefly of agricultural stock, and the free proceeds amounted to £4,932 Scots.

ANDREW HERIOT (son of James Heriot of Trabroun, who died in 1580).

1576-5. Obtained a grant by Deed (dated 27th Nov. 1556, recorded 27th May 1557) of the Lands and Barony of Burnturk, comprehending two parts of the lands of Ballingall, and a quarter of the lands of Glaslee in Fifeshire, from Walter Heriot of Burnturk, the proprietor, under reservation of the liferent of the latter, and subject to a right of redemption in favour of him and his heirs, under the following circumstances :—He (W. H.) having no male heirs, and wishing to conserve his lands to his surname, conveyed them to the above mentioned Andrew Heriot ; but, as he afterwards had an illegitimate son, who was legitimated by the Crown, the lands, according to the contract, reverted to the said Walter. This transaction would seem to point to a blood-relationship between the families of Trabroun and Burnturk.[2]

---

[2] The principal families of the name in Scotland were two, viz.:—that of Trabroun, Berwickshire (1423), and that of Burnturk and Ramorney, Fifeshire. The founder of the Fifeshire Heriots was Walter Heriot, a burgess of Cupar in Fife. He had Charters under the Great Seal of the lands of Ballingall, 1489, Burnturk 1501, and the King's lands of Ramorney, erected with other lands into a Barony, 1512. The Ramorney Heriots are at present represented by Frederick Lewis Maitland Heriot, Esq., an advocate at the Scottish bar, and Sheriff of the County of Forfar.

His (Andrew's) Will, made because he was about to depart for France on business, is dated at Berwick, 2d September 1585, with a codicil dated 5th June 1587. The Will was written by his own hand, and both of these deeds are recorded at Edinburgh, 13th February 1587–8. He seems to have been unmarried, and, probably, boarded in the house of George Heriot senior (on "the north syde of the hie Street" of Edinburgh), as it appears from the codicil he died there on the same day the codicil was subscribed by a notary (5th June 1587), and among the witnesses to the subscription were George Heriot senior, and his son the Founder.

In the Will he says, " I testifie befoir God and all honest pepill that I am frie of all debt of quhat-sumever soume or soumes, and that I am addebted to nane except of the soume of thrie scoir 13 crounes of the sone and half ane croune quhilk I sall God willing pay to James Maiteland [1] and Richart Heriot [2] at my coming to Pariss according to my obligation."

From the quantities of various kinds of cloth and small wares, forming part of his personal effects, it may be inferred he dealt in such articles. The free proceeds of these effects amounted to £3,260, 11s. 5d. Scots.

He left various legacies, among which were the following :—

To George Heriot, senior, 300 merks, "to be distributed be him amangis his dochteris."

To Elspeth Heriot (G. H's. daughter), "alsmekill blak and alsmekill violet as will be hir twa gounes" also "twa ell of camrage."

To George Heriot, younger, whom he calls his " verry friend," his "lytill sword in remembrance," a pair of his "silk shanks," his "hat of castor," his "blak cloik pasimentit (embroidered)," a pair of "perfumit gluvis of Rome," and a pair of gluvis lynit with taffitie ; also to his wife " one pair of gluvis."

To Margaret Heriot, natural daughter to James Heriot of Trabroun, "alsmekill broun as will be ane goun."

---

[1] James Maitland, grandson of Sir Richard Maitland of Lethington.

[2] Richard Heriot was the nephew of Andrew Heriot, and probably the same person who was a subscriber to the King's Confession of Faith in 1580.

To the College of Edinburgh, if there were a sufficient sum after payment of the legacies, 100 merks.

ALEXANDER HERIOT (son of James Heriot of Trabroun, who died in 1580).

1552-53. In the Edinburgh Dean of Guild accounts there is an acknowledgment of " xl s," for the sale to his feu of Ravelston.

1565. Discharged by his father of annual rents of the lands of Ravelston.   Deed subscribed at Elvingston.   Witnesses, James Heriot junior of Trabroun, Andrew Heriot, germanis, &c.—*Protocol Bk. at Hadn.*

PETER HERIOT (son of James Heriot of Trabroun, who died in 1580).

Was murdered at his house in Leith.   The following is a copy of the record of the trial of the murderer :—

"SLAUGHTER."—" HAMESUCKIN."

1587, August 30, THOMAS BONKLE, Cuitlar.

" Delaitil of airt and pairt of the felloune and crewall slauchter of umqle Peter Heriot, induellar in Leyth, brother germane to James Heriot of Trabroune ; committed be way of Hamesuckin within the toune of Leyth uponne the xxix day of August instant."

" PERSEWAR, JAMES HERIOT of TRABROUNE.   Verdict—The Assyis being purgeit of partiall counsall, chosin, sworne, and admitted, and the said Thomas being accusit be Dittay of the cuming be way of Hamesuckin to the said umqle Petir Heriotis dwelling hous in Leyth, and setting vpoune him foirnent the zett thairof with ane drawin sword quharwith he maist crewallie and schamefullie slew him vpoune sett purpois and provisioune &c. ; ffand, pronounceit, and declairet the said Thomas Bonkle to be ffylit and convict of the slauchter of the said umqle Petir."

" SENTENCE.—The Justice ordanit the said Thomas Boncle to be tane to the toune of Leyth, and thair, at place appointit, his heid and rycht arme to be strukin fra his bodie; and all his movable guidis to be escheit and inbrocht to our Souerane lordis vse, for the said cryme. This dome pronounceit be James Nisbett dempstar of Justice Courtis."—*P.C.T.*

He (P. H.) died intestate, but two Inventories of his personal effects were given up, one (recorded 27th February 1588-89) by John Waldie, designed as son to deceased " Margaret Wardlaw spouse wes to the said vmquhile Petir," and a supplementary one, recorded (11th February 1589-60) by " George Heriot portioner of Colelaw in Lauderdaill, his broder germane." The total free proceeds amounted to £2926, 13s. Scots.

The contents of the Inventories chiefly consist of grain and malt, and there is therefore reason to believe he was a grain merchant and maltster.

GEORGE HERIOT of Collelaw (son of James Heriot of Tra-broun, who died in 1580).

1597. Entered (with his wife Katherine Loutheane) into a contract with John Home of Colden-knowes, for certain quantities of grain to be obtained yearly from the said John Home's lands of Maynes of Sauchheid.—*Orig. in Reg. Ho.*

1601. Served heir to his brother Peter, in an annual rent of 20 merks out of the croft land called Channonis Croft, near Lauder.— *R. of R.*

1602. Served heir to his brother Peter, in an annual rent of 24 merks, out of the Kirk lands of Legerwood.—*R. of R.*

MARGARET HERIOT.[1]

1552. Was married to Thomas Fawsyde of Fawsyde, near Tranent, a
or prior. member of an ancient family, the ruins of whose castle still

[1] Margaret, Elizabeth, Alison, Agnes, Janet, and the two Helens, mentioned on this and the two following pages, were daughters of James Heriot of Trabroun, who died in 1580.

remain. There is a Charter under the Great Seal, dated 14th
October 1552, of the lands of Bogend, etc., to "Thomae Fawsyde de eodem et Margreta Heryot his spouse."

ELIZABETH HERIOT.

1558. Was married to Thomas Hamilton of Priestfield.[1] Contract of
marriage dated 23d June, and recorded 28th October. Witnesses thereto—George Heriot in Longniddry, David Heriot in
Arrois, and others.—*R. of D., v. III., f.* 87.

ALISON HERIOT.

1571. "William Pringle ˉof Torwoodlee, married, in 1571, Alison,
daughter of James Heriot of Trabroun, by whom he had three
sons. He died in 1581, and his wife Alison Heriot,[2] who sur-
vived him, had for her second husband David (John?) Renton
of Baillie."—*Burke's Landed Gentry.*

AGNES HERIOT.

1571. Married Alexander Dalmahoy of that ilk (being his second
Wife). The Dalmahoys were an ancient family, and their
estate was in the parish of Ratho.

JANET HERIOT.

1585. Married to John Borthwick, Proprietor, Ballincrieff.

---

[1] "Thomas Hamilton of Priestfield, first Earl of Haddington, President of
the Court of Session, and for a long period Secretary of State and Prime
Minister to James VI., was another contemporary of Craig. He commenced
his career at the Scottish bar in the year 1587. His father was Sir Thomas
Hamilton of Priestfield, sprung from the ancient and honourable family [of
Hamilton] of Innerwick. His mother was Elizabeth Heriot, daughter of
James Heriot of Trabroun. The name of Heriot is not to be forgotten in the
history of Scottish literature. Agnes Heriot, of the family of Trabroun, was
the mother of our great Buchanan." —*Tytler's Life of Craig.*

[2] Her Will is recorded at Edinburgh, 7th December 1592.

HELEN HERIOT. (1.)

ante 1568. Was married to the celebrated Sir Thomas Craig, Advocate,[1] and, after his decease, to Sir John Arnot of Berswick, Provost of Edinburgh.

HELEN HERIOT. (2.)[2]

ante 1580. Was married to George Home, Proprietor in Gullane. Her father in his Will says of him :—" In consideration of the faithfulness and zele of George Home my son-in-law and of the loving favor that he has borne to me, my hous and barnes," &c., " I leve the said George one hundreth pundis."

VIII. ROBERT HERIOT of Trabroun (Son and Heir of James Heriot of Trabroun, who died in 1618).

Married Elizabeth Dundas, daughter of John Dundas, of Newliston.

1608. Granted a bond for £408 to George Heriot, " Jeweller to the Queen's Majesty," promising to repay the same before Whitsunday 1609, with an annual rent of £42 out of his lands. Bond dated at Tranent 24th September. Witnesses, David Seton, bailie of Tranent; Alexander Hume, indweller, Tranent; John Forsyth, servant to the said Robert Heriot; and Cornelius Marshall, servant to the said George Heriot.—*Orig. in Heriot's Hosp.*

1609. Was surety, with others, for Thomas Fairlie, younger of Colineston, for 1000 merks, payable to Thomas Flucker, surgeon, burgess of Edinburgh. Bond dated 21st June, and recorded 28th February 1609.—*R. of D., v.* 158.

---

[1] Tytler in his *Life of Craig* says:—" Sometime previous to this (1568), Craig had married Helen Heriot, a daughter of the Laird of Trabroun," and adds, "the author of Craig's Life, prefixed to the Treatise *De Feudis*, calls Helen Hariot, *femina lectissima Helena Heriota Comarchi de Trabroun in praefectura Hadintoniae filia.'*"

[2] Dr M'Crie says:—"It was very common at that time to have two children of the same Christian name."—*Life of Knox, Edn. vii., App., p.* 434.

1617. Admitted as a burgess of Haddington along with "William Ramsay, brother to John Viscount of Haddington; Mr Robert Lawsoun, brother-germane to umqle Sir James Lawsoun of Humbie; and Alexr. Cokburne, fear of [successor to] Ormiston." —*Hadn. Burgh Record.*

1620. Died intestate in England upon the   day of August. The inventory of his personal effects was given up by "George Heriot, servitor[1] to the defunct," and recorded at Edinburgh 28th February 1621. It is as follows:—

"Ane kow with the followar twentie merks .	£13	6	8
One quoy of thrie year auld price thairof .	10	0	0
Foure quoyis and ane stot of twa yeir auldis price of the peice ourheid fyve pundis .	25	0	0
Ellevin hogis at 30/ the peice . . .	16	10	0
In vtenceillis and domiceillis by the airschip estimat to	20	0	0
	£84	16	8
"Thair wes awin to the said vmquhill Robert Heriot of Trabroune Be George Heriot in Langniddrie. . .	40	0	0
Be the airis and executouris and procurators of vmquhill James Heriot of Trabroune or be Issobell Maitland his relict or be ather of thame twenty foure bollis victuall twa pairt meill and thrie pairt beir," &c. . .	64	0	0
Total (Scots),	£188	16	8"

ELIZABETH HERIOT (Daughter of James Heriot of Trabroun, who died in 1618).

1595. Was married to James Sandilands,[2] of Calder, afterwards Lord Torphichen. Contract of marriage dated and recorded 1st

---

[1] The word "servitor" in those days often signified clerk, secretary, or man of business.—*M'Crie's Life of Knox.*

[2] James Sandilands was second Lord Torphichen, and was in the decreet of the Scots nobility, dated 1606, placed immediately after Lord Boyd, whose peerage dated 1459. He was twice married, but had issue only by his first wife, Elizabeth Heriot above-mentioned. He died in 1617.

August. She was by charter (23d August) liferented in the Mains of Calder, and there is a charter under the Great Seal, dated 15th February 1600, confirming her liferent.

ANNA HERIOT, Elder, and BARBARA HERIOT, Younger (Daugh_ ters of Robert Heriot of Trabroun, who died in 1620).

1623. Granted a discharge, dated 18th February 1623, with consent of the Right Honorable Sir Richard Cockburn, of Clerkington, Knight, and John Dundas, of Newliston, their curators, in favor of Sir John Hamilton, of Trabroun, Knight, with consent of Sir Andrew Hamilton, of Redhouse, Knight, one of the Senators of the College of Justice ; Sir James Foulis, of Colinton, Knight ; Patrick Hamilton, of Preston ; and George Foulis, master cunzeor, curators of the said Sir John, for payment of money due them (the said Anna and Barbara) by the said Sir John, according to contract, the money having been the balance of the purchase-money of the estate of Trabroun, sold under redemption by their grandfather, James Heriot, in 1611.—*R. of D., v.* 338.

ANNA HERIOT (above mentioned).

Married James Brown of Coalstoun, the representative of an ancient family in East Lothian.

1630. Served heiress to Richard Heriot (lawful son of the deceased James Heriot, of Trabroun), her paternal uncle.—*R. of R.*

1631. Discharged George Viscount of Dipling of the sum of 2000 merks borrowed by Laurence Lord Oliphant from Andrew Heriot, her granduncle, and secured over the lands of Dipling.

This sum was inherited by Anna from her grandfather James, who is stated to have been heir of his brother Andrew. The discharge is stated to have been granted by Anna, with consent of her husband James Brown of Coalstoun, Elizabeth Dundas, her mother, and of Catherine Loutheane, relict of George Heriot, brother of Andrew, dated 4th March, and recorded 26th April 1631.—*R. of D., v.* 441.

1633. Served heiress to James Heriot of Trabroun, her great-grand-father, in a third part of the lands of Collelaw, within the baillery of Lauderdale.—*R. of R.*

NOTE.

From the preceding pages it will be seen that the Heriots of Trabroun were a family of eminence in Scotland, and possessed Trabroun and other lands for two hundred years—from 1423 to 1623. In 1611 James Heriot sold Trabroun, with consent of his wife Isabella Maitland (who was liferented in it), to his relative John Hamilton, a member of the Haddington family, under a right of redemption. But as neither James Heriot, nor his son and heir Robert, who died in 1620, redeemed the property, the balance of the purchase-money was paid in 1623 to Anna and Barbara Heriot, daughters of Robert, as per their discharge noted on the previous page ; and Trabroun thus passed entirely from the family. It is now owned partly by the Earl of Lauderdale and partly by Captain Allan.

The direct male line of the family appears to have become extinct by the death of Robert Heriot in 1620, and at present it is uncertain who is the nearest male representative.

IX.—REGULATIONS OF GEORGE HERIOT'S HOSPITAL SCHOOLS.

Chap. I.—Election, Allocation, and Admission of Children.

1. Those eligible for election as pupils are, 1*st*, The children in poor circumstances of deceased Burgesses and Freemen of Edinburgh ; 2*d*, The children of such Burgesses and Freemen as are not sufficiently able to maintain them ; 3*d*, The children of poor citizens or inhabitants of Edinburgh.

2. None of the last mentioned class shall be chosen as long as there are applications for admission on behalf of either of the other classes.

3. Whilst children belonging to the first and second classes shall be received irrespective of their places of residence, those of the third class must at the time of application be resident, and have been resident for one year, within the City.

4. Printed Schedules or Petitions for Admission shall be given out at the office of the Treasurer to the Hospital; and as soon as these have been duly filled up they shall be returned to the same office.

5. The Education and Schools Committee, or an Admission Committee forming, or a Sub-Committee of their number, shall hold meetings at such times as may be deemed necessary for the examination of Petitions ; and when satisfied that the children are eligible, the Convener or some member of the Committee shall indicate on the Petition the School to which the child is allocated, and shall add his initials to authenticate the same.

6. The Petitions so sanctioned shall be delivered to the Treasurer, in order that he may forward the same to the Head Masters of the respective Schools, who shall receive and enrol them as vacancies occur in the School.

7. On every Petition for admission there shall be printed the following Regulations, which are the conditions of admission to the Schools :—

(1) No family receiving Parochial Relief need apply, because by Act of Parliament the children of all such are required to be educated by the relieving parish.

(2) Children admitted into the School shall be allowed to continue during the pleasure of the Governors, and while they behave properly.

(3) Should any of the Boys or Girls be absent from School without a proper cause, and without the Teacher's leave, they shall for the first offence be admonished by the Teacher ; for the second offence they shall be sent home and the case reported to the Inspector, whose consent

must be obtained before they can return to the School; and for the third offence they shall be expelled and their places immediately filled up.

(4) Personal cleanliness on the part of the Children is *at all times indispensable,* and unless this regulation be *strictly observed* the Teacher will find it necessary to send the offenders home.

(5) Hours of attendance *precisely* from 9 to 12, and 1 to 3 o'clock. Infant Schools 10 to 12, and 1 to 3.

(6) The School privileges are only for the children of the Parents residing within the City, except in case of children of Burgesses and Freemen.

CHAP. II.—EDUCATION.

8. The branches taught in the Juvenile Schools shall be English, Reading, Spelling, Grammar, Geography (including the use of the Globes), Writing, Arithmetic, Linear Drawing, the first principles of Mechanics, Physical Science, Natural History, and Vocal Music.

9. In addition to these branches, the Girls shall receive daily an hour's instruction from the Sewing Mistress in Sewing and Knitting; and an hour's instruction three times a week in Cutting, Shaping, and Fitting.

10. The attention of the Teachers shall at all times be directed to the religious and moral training of the Children, as the special object aimed at by the Foundation.

CHAP. III.—ORDINARY INSPECTION TO BE EXERCISED BY THE GOVERNORS, AND ANNUAL EXAMINATION OF THE SCHOOLS.

11. The immediate superintendence of the Schools shall be vested in the Education and Schools Committee.

12. It is earnestly recommended that each Governor visit the Schools as often as his convenience may admit, and make

e

himself thoroughly acquainted with the details of their administration, without any direct interference with the discipline maintained or the mode of instruction pursued.

13. The Governors shall also, in rotation, as indicated in the printed List of Committees, visit the Schools; and in the Minute-Book kept in each School may record, if they see fit, what comes under their notice.

14. Besides the inspection of the Schools generally by the Governors, the Treasurer shall endeavour to make arrangements with two or three Governors for taking a more special interest in the School most convenient for them, by visiting it as frequently as possible, and reporting to him for the information of the Education Committee anything which it is important they should be made acquainted with.

15. The annual examination of the Schools shall take place in the last week of July, on a day or days to be fixed by the Governors or their Education Committee.

16. The Governors shall be allocated as Examiners to the different Schools; one of them being appointed to preside, who shall insert in the School Minute-Book an account of the proceedings of the day.

CHAP. IV.—REWARDS AND PUNISHMENTS,—ANNUAL VACATION, AND OCCASIONAL HOLIDAYS.

17. Such Rewards as the Governors may from time to time sanction shall be bestowed upon the Children, not only for proficiency in their studies and needlework, but also for cleanliness, and general good conduct.

18. The Prizes, furnished at the expense of the Hospital, shall be selected by the Masters and Mistresses respectively, and awarded according to their decision.

19. Corporal punishment shall always be administered in a suitable and temperate manner, and on no account by Pupil Teachers. Punishment with a pointer or other hard instrument is strictly prohibited.

20. In matters of ordinary discipline, after repeated remonstrances the Head Teacher shall suspend the offender, intimate the suspension to the parents or guardians, and immediately report to the Inspector, whose consent must be obtained before his or her re-admission to the School. Every such case shall be reported by the Inspector to the Education Committee.

21. The Vacation shall take place immediately after the Annual Examination of the Schools, and the Children shall re-assemble on the first Monday of September.

22. In addition to the Vacation, the following occasional Holidays shall be allowed :—From the Thursday to the Monday (inclusive) of the Communion Seasons in April and October,— from Christmas day to the day after New Year's day (inclusive), —the Queen's Birth-day,—George Heriot's Day, and the day after.

CHAP. V.—THE TREASURER.

23. He shall, as often as circumstances admit, visit the different Schools, and communicate with the Inspector when he observes anything wrong.

24. He shall, upon the written requisitions of the Masters and Mistresses, and provided he sees no objections to what is asked, supply from the store kept at his office such books, stationery, sewing, and other materials, as the Governors allow for the Schools ; he shall see that all these are supplied by means of a pass-book for each School ; and when he shall consider any demand unnecessary, he shall consult the Education Committee before complying with it. At the end of each year he shall prepare and lay before the Education Committee an abstract of the different articles furnished to each School.

25. He shall take care that, once a month, each Head-Master furnish him with a statement, to be laid before the Governors, of the average attendance of pupils on the roll at his School for the whole of the past month ; also the number who have left, and the names of any who have been suspended or dismissed since the previous report. He shall submit such

statements to the Admission Committee, whose duty it is to allocate to the different Schools the children they admit, with a due regard to the number of vacancies and, as far as possible, the convenience of the scholars. He shall furnish each Teacher with the names of the children thus allocated to his or her School, no different allocation and no transference of scholars from one school to another being on any account allowed, except with the consent of the Admission Committee.

26. He shall keep a Book (each School forming a head in it as in a ledger), accessible to the Governors and the Inspector of the Schools, containing the names of the Teachers, Assistants, and Pupil Teachers in each School, the dates of their appointment, their salaries or wages, and their places of residence ; and each Teacher shall be bound to report immediately to the Treasurer every change of Assistant, the reason for such change, and the name and residence of any new Assistant temporarily appointed, in order that such information may be entered in this book. He shall also briefly record in it any facts and circumstances which may come to his knowledge bearing upon the working of each School, in order that a condensed history of the School may be preserved for reference.

Chap. VI.—The Inspector.

27. He shall take a general and prudent oversight of the Schools, and be the ordinary organ of communication with the Governors, through the Education Committee, in regard to them.

28. He shall specially give such attendance as may enable him to afford evidence to the Governors that the prescribed hours of teaching are punctually observed, and such educational results obtained as might reasonably be expected ; he shall report any irregularities or defects in the working of the Schools to the Education Committee, through the Treasurer, from time to time ; and when he has occasion to suggest anything relating to the wellbeing or carrying on of any of the Schools or its apparatus, or books, &c., he shall do so to the Education Committee through the Treasurer of the Hospital.

29. He shall, along with the Treasurer, advise with the Teacher on matters of ordinary discipline as well as on all questions which may arise between them and their Assistants or Scholars ; he shall arrange with the Teachers the best mode of making known to parents all cases of absence from school without proper cause or leave, and of carrying into effect Rule 7/3 of the Regulations referring to children's exclusion from school ; he shall immediately be made acquainted with every case of suspension for absence or any other cause, and all such cases shall be regularly reported by him to the Education Committee. He shall consult the Education Committee on any special case seeming to demand further discipline, and no expulsion from the Schools shall take place without their knowledge and consent.

30. He shall see that the different Head-Teachers keep an account of the number of the pupils in attendance each day, and that the School Registers, comprehending the names of the Head-Master, Sewing Mistress, Singing Master, and all the Assistants and Pupil Teachers, regularly record all cases of absence or lateness, and the extent of each ; he shall take care that these daily Accounts and Registers, initialed each day by the Head-Master, are submitted to him every visit, that he may examine and initial them ; and state how far he finds them to correspond with the results of his personal examination.

31. In the event of the absence of a Teacher, Assistant, or Pupil Teacher, from illness or any other cause, he shall see to their places being temporarily filled by some one competent to discharge the duties required, and he shall see that every instance of absence has been immediately notified to the Treasurer as well as to himself by the Head-Master or Teacher in charge, and he shall report to the Education Committee every instance where this is not done.

32. He shall monthly present to the Education Committee, through the Treasurer, a written report on the Schools which he has visited, including answers to the following Queries :—

1. What Schools have you examined during this month ?

2. On what day and what hour did you visit these Schools ?

3. Did you examine the Attendance Rolls or Registers of the Pupils, and find them correctly kept ?

4. Did you examine the Daily Time Registers of the Teachers, and find them correctly kept ?

5. State, from personal examination, the actual numbers of Pupils on the Roll and in Attendance at the hour of your visit to these Schools, and the numbers on the Roll and in Attendance from the Time-book the day immediately preceding your visit.

6. What was your opinion of the discipline of the Schools you visited, and the general appearance of the Pupils ?

7. What classes did you examine in these Schools, and were you satisfied with their condition ?

8. Have you reason to suppose that any of the School regulations have been disregarded or evaded ?

9. State any other matter to which you think the Governors' attention should be at present directed.

33. He shall, at least twice in the year, ascertain the number of school books and amount of stationery, &c., used in each school during the previous half-year, and shall report the same to the Education Committee.

CHAP. VII.—HEAD TEACHERS, ASSISTANTS, AND PUPIL TEACHERS.

34. The Head Teacher in each School shall keep a School Attendance Book, in which shall be entered every morning and afternoon the exact time of the arrival at school of each Teacher, Assistant, Pupil Teacher, or Monitor; there shall also be entered in this Book the duration and cause of any incidental absences during the teaching hours.

35. It shall be the duty of each Head Teacher to see that the School is daily opened and closed with praise and prayer.

36. The Head Teacher shall teach the highest class in the School, superintend the tuition of the other classes, prescribing their work and the distribution of their time, and appoint to all

the other teachers the class or classes to be specially taken charge of by them.

37. The Head Teacher shall prepare for each class a Time Table of Course of Studies, which shall be rigidly adhered to, in order that no essential subject of instruction shall be neglected, and also that evidence be afforded to the Governors that the time is fairly distributed among the various subjects taught. These time-tables are to be hung up in the different Class-Rooms or attached to the different Roll-Books, and it shall be one of the duties of the Inspector to ascertain whether or not they are acted upon.

38. The Head Teachers shall from time to time, as they see occasion, re-arrange the divisions of their Schools, in order that the more apt and diligent of the Pupils may be advanced to a higher division.

39. They shall direct the Assistant Teachers how to conduct their different Divisions ; and have the power of requiring their attendance for a reasonable time before and after School hours, to assist in what they consider the necessary work of the School. They shall be particularly careful to see that the Assistants are in attendance ten minutes before the regular hour of meeting ; they shall likewise mark the Time-Book themselves, and see that it is marked by the other parties in their Schools.

40. The Head Teacher shall arrange with the Sewing-Mistress as to the hour at which each Class shall be taught in her room.

41. In the School Register, according to the date of admission, the Head Teachers shall insert and number consecutively the names and ages of the pupils, the names of the parents or guardians, with their places of residence, the result of their examination, and the class to which each pupil is assigned. This Register shall be neatly and accurately kept by the Head Teacher, posted up monthly, and transmitted in a complete state to the Treasurer at the close of each Session in July, for the inspection of the Governors. After the vacation it shall be returned to the Teacher.

42. The Head Teachers shall obtain by means of Pass-Books from the Treasurer all the Books and other materials required and provided for the Schools, and they shall not only send these Pass-Books along with separate order lists, but they shall also keep detailed accounts of the supplies furnished to them under their proper dates, which accounts shall be submitted to the Inspector for examination whenever required, and at least once every half-year.

43. The Head Teachers shall not only have full discretionary power regarding the arrangements and general management of the Schools, but shall be expected specially to attend to the cleanliness and proper deportment of the Pupils, and to employ every means to promote good order in every department. They shall see that the property is not damaged, and shall report to the Superintendent' of Works in writing all breakages and damage to furniture or building so soon as observed. They shall also see that the Books, Stationery, and other educational apparatus are kept in proper condition, and that no waste of any kind is allowed either by the Scholars or Assistant or Pupil Teachers.

44. The Head Teachers shall see that their respective Schools and all the conveniences are kept at all times properly clean and ventilated ; and they shall report to the Superintendent of Works if the Cleaners have not attended to their duties in the cleaning or ventilation of the Schools, as provided for in Article 95.

45. The staff of Assistant and Pupil Teachers for each School shall be fixed and determined by the Education Committee from time to time as circumstances may seem to them to require this to be done.

46. All Assistants shall be elected by, and shall hold their appointment during the pleasure of, the Governors or their Education Committee.

47. The Assistants shall be under the direction of the Head Master as to their time and their work, as already specified (see Art. 36). They shall devote themselves anxiously to the daily preparation of the several subjects on which they require to examine the Classes.

48. The Female Assistant Teachers and Pupil Teachers shall be entitled to receive weekly two hours' instruction from the Sewing Mistress.

49. The Assistants shall be responsible to the Head Master or Head Mistress for the progress of the Children under their charge in the different branches prescribed.

50. The Head Teachers of the different Schools shall be entitled to recommend to the Education Committee from time to time such young persons as they would desire as apprentices; and if the Committee shall be satisfied that any so recommended are, after examination by the Inspector, likely to prove efficient Teachers, their names shall be entered upon the list of Candidates kept by the Treasurer, but no appointment shall be made without the sanction of the Education Committee or its Convener.

51. Any Candidate or Monitor taken as a Pupil Teacher to qualify for becoming a Certificated Teacher shall be not less than 13 years of age complete at the date of the commencement of the Apprenticeship, which shall be from the 1st of the month on which the Scotch Education Committee of the Privy Council shall hold the school year to commence, and succeeding the examination which the Candidate shall have successfully passed, and become entitled to be a Pupil Teacher under the Code.

Note.—The age may vary according to the requirements of the Code issued from time to time by the Education Department.

52. The Apprenticeship shall be for five years, or for such other period as shall correspond with the Report of Her Majesty's Inspector of the Examination the Applicant has passed successfully, and the Applicant shall be bound to enter into Indentures in the terms required of the Education Department under the Code at the time in force.

53. The Head Teacher of every School (Juvenile or Infant) shall give not less than one hour's instruction every week day to his or her own Apprentices in the art of Teaching and the various branches taught in the School or required by the Government Regulations for the respective years of the Apprenticeship (Latin and French excepted), and shall be bound to do

everything in his power to prepare the Pupil Teachers for their Examinations. In the event of the Education Department at any time requiring that more than one hour daily be devoted to the Instruction of the Pupil Teachers, then such additional instruction shall be given as if it had been specially mentioned in this article.

54. The Apprentices or Pupil Teachers, in addition to giving their services as Teachers during School hours, shall regularly attend the class or instructions of the Head Teacher as required under the Code issued by the Education Department, and they shall further attend such classes or instructions as the Governors or their Education Committee may from time to time direct, or which may be required to prepare them for their Examinations.

55. All Apprentices or Pupil Teachers shall be required to appear annually before the Government Inspector (or such other party as may be appointed by the Governors or the Education Committee) for Examination upon the subjects prescribed by the Education Department as applicable to the year of Apprenticeship or study they are then closing, and in the event of their failing to pass the said Examination satisfactorily, it shall be in the power of the Governors to terminate the engagement, or to restrict the salary to be paid for the next year.

56. The salary or stipend payable to Apprentice or Pupil Teachers shall be :—For Male Apprentices—for the first year they serve the Governors as Apprentice, £15 ; for the second year, £17, 10s. ; for the third year, £20 ; for the fourth year, £22, 10s. ; and for the fifth year, £25. And for Female Apprentices :—For the first year they serve as Apprentices to the Governors, £12, 10s. ; for the second year, £15 ; for the third year, £17, 10s. ; for the fourth year, £20; and for the fifth year, £22, 10s. The said sums shall be payable by quarterly instalments.

57. In the event of a Pupil Teacher leaving in the course of a year, the vacancy shall be temporarily filled up by the Education Committee till the next Examination of Pupil Teachers, by a Monitor, Candidate, or Probationer, who shall be paid at such

rate per annum as may be fixed by the Governors or their Education Committee.

58. It shall be the care of each Head Teacher so to arrange his Apprentices or Pupil Teachers that they may be in different years of their Apprenticeship, so as to equalise the payments, and to avoid the inconvenience of having more than one in the first year of Apprenticeship in any School at one time.

59. The Head Master shall without delay report in writing to the Treasurer every change in the staff of the School, whether caused by resignation or otherwise, giving the name of the party so leaving, with the date and reason thereof, and the name of the party supplying the vacancy, whether temporarily or permanently.

CHAP. VIII.—SEWING MISTRESSES.

60. The Sewing Mistress shall be present daily at the Devotional Exercises at the opening and closing of the School.

61. She shall give each Class of Girls one hour's instruction daily in Sewing and Knitting ; and she may allow the girls to do Fancy Work, but only when they are satisfactorily advanced in plain Sewing, Knitting, and Mending. She shall also take a special charge of the Girls in regard to cleanly and tidy habits.

62. The Mistress shall assist in such arrangements as the Master may find necessary for securing order and quietness in the Classes entering or leaving her Department. She shall also give such attendance during the hours for instruction of the Female Pupil Teachers as may from time to time be required by the Education Committee.

63. The whole materials for the Sewing Department shall be received through the Treasurer's Office by an order from the Mistress accompanied by a pass-book. The Sewing Mistress shall also attend to Regulation No. 42 in regard to matters falling under her department. *The materials required for any Fancy Work must be procured by the pupils.*

64. The Mistress shall teach Cutting, Shaping, and Fitting during one hour at the close of the ordinary lessons for the day

on three days each week, and until the pupils acquire sufficient proficiency to render it advisable to entrust them with material such as Cotton or Linen, they shall be supplied with paper upon which to be instructed.

65. The Mistresses shall see in the Schools where Sewing Machines have been introduced that the Senior Female Scholars to be instructed in the use of them shall be those who in their opinion are proficient in Hand Sewing, so that instructions in the use of the Machine may be looked upon as a reward for proficiency in Hand Sewing.

CHAP. IX.—TEACHERS AND ASSISTANTS OF INFANT SCHOOLS.

66. No children shall be admitted to the Infant Schools under five years of age.

67. Devotional Exercises suited to the capacities of the Children shall be conducted daily at the opening and close of the School.

68. Bible instruction shall be given every day, and also a lesson on some interesting object in Natural History.

69. The Teacher shall constantly keep in view, in whatever lessons are taught, and in the collateral information communicated to the Children, the gradual development of their mental and moral powers ; and shall endeavour to prepare them for entering with advantage a Juvenile School. She shall also prepare the Children for passing the necessary examinations by the Government Inspector.

70. The business of the School shall be enlivened from time to time by the singing of Hymns and Moral Songs; and a little relaxation shall be allowed to the children at the end of each hour.

71. The Infant School Teachers shall take especial care to enforce the Regulations as to the cleanliness and regular attendance of the Children.

72. The Assistants and Pupil Teachers in the Infant Schools

shall take their instructions for the discharge of their ordinary duties from the Principal Teacher; and shall be expected to perform such extra duties as may from time to time be required for the benefit of the Children.

73. The Principal Teachers of the Infant Schools shall regularly keep the roll of their pupils, and the School Register, in the same manner as provided for the Juvenile Schools in Article 41. They shall also attend punctually to the keeping of the Time-Book, and attend to such other regulations in Chapter VII. as may be applicable to their department.

CHAP. X.—TEACHERS OF VOCAL MUSIC.

74. The Teachers of Music shall devote to the Pupils and Pupil Teachers in each School such time as the Education Committee may from time to time appoint.

75. In their instructions the Teachers shall endeavour to give their pupils an accurate acquaintance with the principles of Music, as well as train them in singing.

76. The ordinary lessons of Psalmody may be varied by short Anthems, good National Songs, or other simple pieces of music, adapted to such words as may inculcate morality or elevate the feelings.

CHAP. XI.—REGULATIONS IN CASE OF ABSENCE FROM ILLNESS OR OTHERWISE.

1. *Head Teachers and Sewing Mistresses.*

77. If any of these become unfit from illness for duty, they shall immediately report to the Treasurer and Inspector. If the illness is of such a nature as to render it probable that they will be absent from duty more than three days, a medical certificate to that effect shall be sent to the Inspector, who shall immediately communicate with the Treasurer and the Convener of the Education Committee, with the view, if thought necessary, of providing a substitute. Should any substitute be appointed for a Head Master, he shall be paid at the rate of   a-week, and for a Sewing Mistress, she shall be paid at the rate of   a-week.

The Inspector shall report at the next Meeting of the Education Committee what means were adopted during every such case of absence.

2. *Assistants and Pupil Teachers.*

78. An absence of three days for an ordinary illness may be allowed by the Head Master, provided that the Assistant or Pupil Teacher give satisfactory evidence to him that such absence is caused by sickness, and that he reports the same at once to the Treasurer and Inspector.

79. If the illness should prove more serious than was at first anticipated, and shall extend beyond a week, application, accompanied by a medical certificate, shall be made to the Treasurer for a fortnight's further absence, and should a paid substitute be required during this absence the payment shall be made by the Governors. If, however, at the outset the illness should threaten to extend over a longer period than a week, application shall be made in the first instance for a three weeks' absence, the Governors defraying all necessary expenses in providing a supply. All such cases shall be reported to the first meeting of the Education Committee.

80. If the illness shall extend beyond three weeks, fresh application shall be made to the Education Committee through the Treasurer for a further leave of absence not exceeding five additional weeks, on the understanding that one-half of the payment made to the substitute for the said additional period shall be deducted from the salary of the absent Teacher.

81. If, however, any Assistant or Pupil Teacher shall be absent from sickness for two entire months, the Inspector is required to make a special report of the case to the Education Committee, in order that they may consider whether it is desirable that the teacher's connection with the School should be closed.

82. The allowance made to any substitute for an absent Assistant shall be   . It must be distinctly understood that any temporary appointment does not imply or give any claim to any more permanent appointment under the Governors.

83. Each Head Teacher shall report immediately to the Treasurer the name and residence of every Assistant temporarily appointed, in order that such information may be entered in the Book containing the names of the Teachers and Assistants.

84. The Masters and Mistresses of the different Schools shall keep a Pass-Book, in which the names of their different Assistant Teachers, with the weekly wages of each, and every instance of absence for a day or half a day without leave, as above, shall be entered, and this Pass-Book shall be sent to the Treasurer before the end of each quarter, to enable him to check what is due to each.

CHAP. XII.—THE SUPERINTENDENT OF WORKS.

85. He shall have the superintendence of the fabric of the different Schools, as well as of their furniture, and take care that the whole is kept in proper repair.

86. He shall also have the charge of all the Cleaners, and see that they do their duty properly. He shall also have, subject to the approval of the Treasurer, the power of appointing and dismissing Cleaners, and making all necessary interim arrangements, but all such appointments and dismissals and interim arrangements shall be reported to the Education Committee for their sanction. He shall also order and regulate the necessary supply of coals and gas.

87. He shall see that every Cleaner is furnished with a book containing regulations for her guidance, and that there are recorded in it, for her information, all grants by the Governors of the use of any of the School-rooms, with the conditions of such grants, and any recall of grants; and he shall see that the conditions are adhered to, that the Cleaners on no pretence whatever allow the School-rooms to be used by any other persons; and that they report to him immediately any damage done to the Schools, or any violation of the conditions of their use.

88. He alone shall order repairs and (after receiving the Governors' instructions) any alterations or additions to the Schools or furniture, and these, together with all supplies of

coal, he shall make by written orders; and all persons entitled to visit the Schools having any suggestions to make on the matters thus placed under his charge, shall make them in writing to the Treasurer.

CHAP. XIII.—CLEANERS OF THE SCHOOLS.

89. Each Cleaner shall be in attendance when the School is opened, and also when it is dismissed, to take charge of the rooms, which she shall carefully sweep and dust every day.

90. She shall attend to the ventilation of all the rooms, and particularly to everything in and around the School which may be essential to comfort and cleanliness.

91. She shall put the fires on at the proper time, and supply the scuttles with coal; she shall be careful in putting out the fires and gas, and shall also take care that there is no waste of either.

92. She shall enter in a book to be kept by her the number of fires daily put on in each School, and this book shall be brought by her to the Superintendent of Works once a month, that he may examine the same and report thereon if necessary to the Education Committee.

93. She shall wash all the Class-rooms, Stairs, and premises at least once a fortnight.

94. She shall take her directions for the discharge of her duties from the Superintendent of Works.

95. The Head Teacher of each School shall see daily that the School and premises are properly cleaned and ventilated, and if this is neglected he shall direct the Cleaner to have it done, and if this direction is not at once attended to he shall forthwith report the same to the Superintendent of Works.

INDEX.

.

www.ingramcontent.com/pod-product-compliance
Lightning Source LLC
Chambersburg PA
CBHW021953190326
41519CB00009B/1244